Steel–concrete composite bridges

David Collings

Thomas Telford

Published by Thomas Telford Publishing, Thomas Telford Ltd,
1 Heron Quay, London E14 4JD.
www.thomastelford.com

Distributors for Thomas Telford books are
USA: ASCE Press, 1801 Alexander Bell Drive, Reston, VA 20191-4400, USA
Japan: Maruzen Co. Ltd, Book Department, 3–10 Nihonbashi 2-chome,
Chuo-ku, Tokyo 103
Australia: DA Books and Journals, 648 Whitehorse Road, Mitcham 3132,
Victoria

First published 2005

Figures 5.2, 10.10, 11.2 courtesy of Benaim
Figure 10.9 courtesy of T Hambly
Figures 5.4, 5.5 courtesy of Amec Group Ltd
Figure 10.12 courtesy of Arup.

A catalogue record for this book is available from the British Library

ISBN: 0 7277 3342 7

Typeset by Academic + Technical, Bristol
Printed and bound in Great Britain by MPG Books, Bodmin

Dedication

For my father

Contents

Foreword

The bridge crossing it, with its numberless short spans and lack of bigness, beauty and romance he gazed upon in instant distain. It appeared to creep, cringing and apologetic, across the wide waters which felt the humiliation of its presence . . . Yet he received a shock of elation as the train had moved slowly along the bridge, carrying him with it, and he gazed downward upon flowing waters, again he marvelled at what men could do; at the power of men to build; to build a bridge so strong . . . [1].

I see this book as a journey. A journey of experience from the first simple river crossing to the more complex suspended spans of the early twenty-first century. A journey across the world from the bleak post-industrial landscapes that are still scattered across Britain, around the broad untamed rivers of Bengal and into the racing development of South East Asia. But it is also a subjective journey, over and under the numberless spans of motorway bridges that are the 'bread and butter' of many bridge designers, through to the countless bridges that perform their task with pride, and always marvelling at how we build so strong, always questioning.

This book has its origins in the composite bridge chapter of the *Manual of Bridge Engineering* [139]. This book expands upon that chapter and provides details of more steel–concrete composite bridges. It is intended to show how composite bridges may be designed simply from basic concepts without the need for a clause-by-clause checking of codes and standards. All chapters use examples of various bridges to illustrate the design and construction methods. The book looks impartially at this construction form and compares composite bridges with other types, and often places limits on their use. The book is intended for a number of readers, first those who use the *Manual of Bridge Engineering* and wish to find more detail on steel–concrete composite bridges. Second, it is for those engaged in design who require a deeper understanding of the methods used as well as how they are verified against design codes. The book aims to show how to choose the bridge form, and design element sizes to enable drawings to be produced. The book covers a wide range of examples, in all of which the author has had an involvement or interest.

Acknowledgements

Many of the examples and photographs are derived from work carried out at Benaim, and the support of the staff and directors is gratefully acknowledged. Particular thanks go to Rup Shandu for his help with drafting some of the more complex figures and to Robert Henderson for patiently reading the drafts of each chapter.

I would like to thank John Bowes and Phil Girling for the aerial views of the Doncaster Viaduct and Brian Bell for the photographs of the Irish bridges, not all of which were used in the final version. I am also indebted to Naeem Hussain and Steve Kite of Arup for the information on the Stonecutters Bridge tower, and to Sally Sunderland for the information on Brunel's Paddington Bridge.

Notation

A	area
A_a	steel area
A_{ac}	composite section
A_c	concrete area
A_s	reinforcing steel area
B	width
D	depth of girder, rigidity
E	Young's modulus
E_a	steel modulus of elasticity
E_c	concrete modulus of elasticity
$E_{c'}$	long-term concrete modulus
F	force or load
F_F	force on fixed bearings
G	permanent load, shear modulus
H	height
I	second moment of area
J	torsional constant
J_m	mass moment of inertia
K	stiffness of member, soil pressure coefficient
L	length of beam or slab
L_e	effective length
M	moment
M_D	design resistance moment
M_f	design resistance moment of flanges
M_u	ultimate moment
M'	reduction of moment
N	axial load
N_D	design resistance load
No	number of connectors
N_{pk}	chord plastification-k joint
N_{px}	chord plastification-x joint
N_u	ultimate axial resistance
N_{ul}	squash load
P	load on single connector, bolt capacity, wind susceptibility factor
P_i	initial prestress
P_o	prestress force after losses
P_u	resistance of a connector
P_{UL}	resistance of a connector in lightweight concrete

Q	variable load
Q_l	longitudinal shear
R	reaction
S	Strouhal number
T	torsion, tension, period
V	shear force, wind speed
V_{arch}	punching shear resistance
V_{cf}	critical wind speed-flutter
V_{cv}	critical wind speed-vortex shedding
V_D	design shear resistance
V_w	design shear resistance of web only
W	load
Z	section modulus
Z_b	bottom flange modulus
Z_c	modulus of concrete element
Z_t	top flange modulus
a	web panel length
b	width of section
d	depth
e	eccentricity
f	stress, frequency
f_{ck}	concrete cylinder strength
f_{cu}	concrete cube strength
f_u	ultimate tensile strength
f_y	yield strength
g	gap
h	height of section
i	integer
k	coefficient, constant
m	mass per length unit
m_f	relative flange stiffness
n	modular ratio
r	radius of gyration, radius
s_b	spacing of reinforcing bars
t_f	flange thickness
t_w	web thickness
v	shear stress
x	distance along member
y	distance
z	lever arm
α	axial contribution factor, prestress loss
γ_{fl}	partial load factor
γ_{fm}	partial material factor
γ_{f3}	partial factor (BS 5400)
δ	deflection, settlement
ε	strain
θ	angle
λ	slenderness parameter

| ϕ | creep function |
| φ | dynamic increment |

Subscripts

a	steel
a–c	steel–concrete composite
b	bottom
c	concrete
cr	critical buckling
c–s	steel–concrete composite, cracked section
des	destabilising
f	flange
i	integer
m	moment
max	maximum
min	minimum
s	reinforcing steel
t	top
u	ultimate
ua	ultimate strength of steel shear connector
v	shear
w	web
y	yield
C	compression
DW	distortional warping
N	axial effects
T	tension
TW	torsional warping
0,1,2	general number, construction stage

General concepts

...the composite whole being substantially stronger and stiffer than the sum of the parts...

Introduction

Composite bridges are structures that combine materials such as steel, concrete, timber or masonry in any combination. In common usage nowadays composite construction is normally taken to mean either steel and concrete construction or precast-concrete and insitu-concrete bridges. For this book the scope is limited to steel–concrete composite structures. Composites is also a term used to describe modern materials such as glass- or carbon-reinforced plastics and so on. These materials are becoming more common but are beyond the scope of this book. Steel–concrete composite structures are a common and economical form of construction used in a wide variety of structural types. This book initially reviews the forms of structure in which composite construction is used, and then the more common forms of composite construction are considered in more detail.

Compliance with codes and regulations is necessary in the design of a structure but is not sufficient for the design of an efficient, elegant and economic structure. An understanding of the structure's behaviour, what physically happens and how failure occurs is vital to any designer; without this understanding the mathematical equations are a meaningless set of abstract concepts. It is also vital to understand how the structures are constructed and the effect this can have on the stress distribution. One aim of this book is to give an understanding of the behaviour of composite structures. Examples of designs of composite structures are used extensively throughout the book. Where possible these examples are based on designs or checks carried out by the author.

Structural forms

Most commonly, steel–concrete composite structures take a simple beam and slab form. However, composite structures are very versatile and can be used for a considerable range of structures – from foundations, substructures [10, 11] and superstructures through a range of forms from beams, columns, towers and arches, and also for a diverse range of bridge structures from tunnels [12], viaducts [16], elegant footbridges and major cable-stayed bridges.

Steel–concrete composite bridges generally occupy the middle ground between concrete and steel structures; they are competitive with concrete bridges from spans of about 20 m in basic beam and slab forms. For heavier loads, as on railways, deeper through-girders or truss forms are more likely. From 50 m to 500 m steel–concrete composite arches and cable-stayed bridges are competitive.

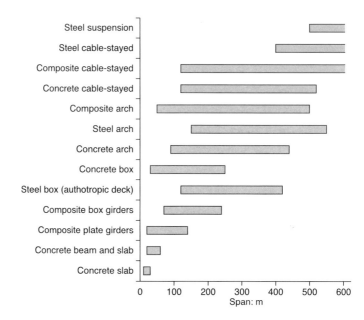

Figure 1.1 Span ranges for various bridge types.

For the longer span bridges, lighter all-steel structures are usually preferred. Figure 1.1 illustrates the typical span ranges for the more common bridge forms.

Materials

The behaviour of the composite structure is heavily influenced by the properties of its component materials. The reader wanting to understand composite bridges should first have a good understanding of the properties and design methods for the individual materials. In particular the reader should note the differences between materials, as it is the exploitation of these different properties that makes composite construction economic. Concrete has a density of approximately $25\,kN/m^3$, a compressive strength of 30 to $100\,N/mm^2$ and almost no tensile strength. Steel has a density of $77\,kN/m^3$, a tensile strength of 250 to $1880\,N/mm^2$ and is prone to buckling where thin sections are loaded in compression. The use of a concrete slab on a steel girder uses the strength of concrete in compression and the high tensile strength of steel to overall advantage.

Codes

Bridges are designed to conform to certain standards outlined in codes of practice. In the UK the principal bridge code is BS 5400 [2, 3, 4, 5], this is supplemented by a series of documents by the Highways Agency or Network Rail. Standards in other countries are also based on these British Standards [21, 52]. European standards [7, 8, 9] are being introduced and will eventually supersede the British Standards. Engineers designing structures abroad may well use other codes such as the American AASHTO standard [22]. The physical behaviour of the composite bridge is the same irrespective of the design standard; an understanding of the behaviour of the structure will allow the designer flexibility in adapting to whichever code is required for a particular design.

Most modern codes are based on similar limit state principles where the designer has to consider two broad states: the ultimate limit state, where the various failure modes are considered; and the serviceability limit state, where cracking, deflection, durability, vibration and other criteria are considered. Modern codes are often bulky documents and the clause-by-clause checking of each element for each limit state and all possible load combinations would be a daunting, tedious and wasteful task. The designer has to make decisions as to what will be designed and what will be assumed to comply, with little checking in the way of detailed calculations. In the following chapters, pointers to what requires design and what can be assumed to comply are given.

The design formulae given and examples presented are generally in accordance with British Standards. These standards use the partial limit state format with three primary factors. The partial load factor γ_{fl} is applied to the loads, generally to increase them. The partial material factor γ_{fm} is applied to the material strength, generally to reduce them. The third factor γ_{f3} is slightly anomalous: for concrete structures it is usually applied to the loads with γ_{fl}, whereas for steel structures it is usually applied to materials with γ_{fm}. For steel–concrete composite structures some confusion can occur when both materials are combined; for examples in this book the concrete convention is used.

Concrete

Concrete is a material formed of cement, aggregate and water. The proportions of the components are varied to obtain the required strength (generally the more cement and less water added the stronger the resulting concrete). Admixtures to improve workability, retard strength gain and so on may also be added. The primary property of concrete of interest to the engineer is its compressive strength. To date in the UK, codes of practice have used concrete cubes to determine the ultimate compressive strength (f_{cu}). In the USA and Europe cylinders have been used to determine strength (f_{ck}). In this book we will use both UK and European standards so both f_{cu} and f_{ck} will be used. Generally cylinder strengths are 80 to 85% of cube strengths. In the examples both cube and cylinder strengths are specified.

The tensile strength of concrete is normally ignored for design; however, in some circumstances, that is when looking at the cracking of concrete, it may be useful to have an estimate. The tensile strength of concrete f_{ct} is about 10% of its compressive strength.

$$f_{ct} = 0.1 f_{cu} \tag{1.1a}$$

$$f_{ct} = 0.45 (f_{ck})^{0.5} \tag{1.1b}$$

A stress–strain curve used for the design of concrete elements is shown in Fig. 1.2(a). The ultimate failure strain is typically 0.0035 with a limit of elasticity at a strain of 0.001 to 0.0015 for concrete strengths below 50/60 (f_{ck}/f_{cu}). For higher strengths the ultimate strain will be smaller.

The concrete modulus is required in composite structures to determine the distribution of load between each material. The modulus for concrete (E_c) is summarised in Table 1.1 for various concrete strengths.

When mixed and placed into the structure the concrete is a fluid and will flow to take up the shape of the supporting formwork. As the concrete hydrates and hardens, the chemical reactions occurring give off a significant heat; the heat will cause expansion and any restraint to this expansion (from the steel element) may

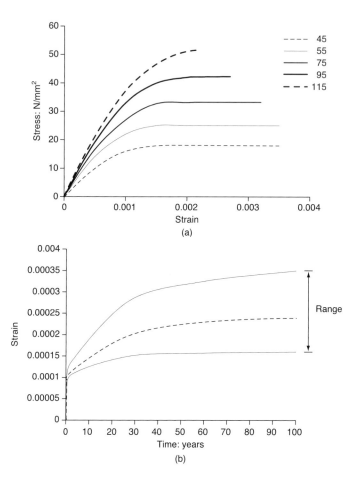

cause cracking. Reinforcement is required to control this early age cracking [23]. For a typical 200 to 250 mm-thick slab of a composite beam bridge, the minimum reinforcement required is approximately:

$$A_s = 0.35 \frac{A_c}{100} \tag{1.2a}$$

This reinforcement is placed approximately equally in the top and bottom faces. If the slab is constructed using an infill-bay system (common for continuous structures, see Chapters 4 and 5) the infill bays will be restrained and require

Table 1.1 Concrete modulus (E_c) for various concrete strengths

Strength f_{ck}/f_{cu}: N/mm^2	Short-term static modulus: kN/mm^2	Modulus range: kN/mm^2
32/40	31	26–36
40/50	34	28–40
50/60	35	30–42
65/80	36	32–43
85/100	38	34–44

4

additional reinforcement:

$$A_s = 0.9 \frac{A_c}{100} \qquad\qquad (1.2b)$$

As the concrete hydrates further over a period of months or years it will shrink slightly. The amount of shrinkage will depend upon the concrete thickness, mix parameters and environmental conditions (primarily humidity). Figure 1.2(b) shows a typical shrinkage strain against time for a 200 to 250 mm-thick slab in the UK. Shrinkage can be important for composite structures, as it will tend to create additional stress at the steel–concrete interface. Design procedures to allow for these stresses are given in the examples in subsequent chapters.

The final property of concrete to be considered is creep. The amount of creep depends upon the magnitude and duration of applied stresses, concrete mix parameters and environmental conditions. The creep affects the concrete modulus; a creep-affected modulus can be calculated from the following equation:

$$E_c' = \frac{E_c}{1 + \phi} \qquad\qquad (1.3)$$

The changing of the modulus will influence the distribution of load between materials. Typical creep factors (ϕ) for common composite components are 1.0 for precast decks to 1.5 for insitu concrete. Generally assuming $E_c' = 0.4E_c$ to $0.5E_c$ will give a reasonable estimate for most circumstances.

For a concrete element such as a deck slab acting as part of a composite structure there are two key design criteria: axial compressive capacity and bending resistance. For a concrete element subjected to a compressive load the ultimate stress in the concrete is limited to $0.4f_{cu}$ or $0.57f_{ck}$ (depending on the code used [3, 7]). The ultimate axial resistance (N_u) of the section is simply the ultimate stress multiplied by the area.

$$N_{uc} = 0.4f_{cu}bd \text{ or } 0.57f_{ck}bd \qquad\qquad (1.4)$$

For concrete elements subject to a bending moment the ultimate moment of resistance (M_u) can be derived by assuming a concrete failure (M_{uc}) or a failure in tension of the embedded reinforcing steel (M_{us}), as in Fig. 1.3.

For failure of the concrete:

$$M_{uc} = Cz$$

assuming the limiting value of compression C occurs when $0.8x = d/2$ and $z = 0.75d$;

$$M_{uc} = 0.15bd^2 f_{cu} \text{ or } 0.21bd^2 f_{ck} \qquad\qquad (1.5)$$

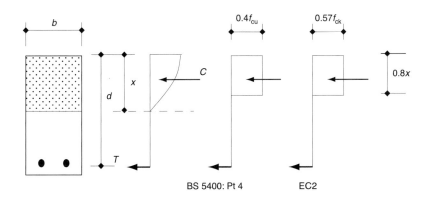

Figure 1.3 Idealised bending stresses in a concrete element.

5

For failure of the steel reinforcement:

$$M_{us} = Tz, \quad T = 0.87f_y \quad \text{and} \quad z = 0.75d$$

$$M_{us} = 0.65dA_sf_y \tag{1.6}$$

To achieve ductility it is normal to ensure that the steel fails before the concrete and that M_{uc} is greater than M_{us}. The concrete may also fail in shear, direct tension or by a combination of axial force and moment. These other less common design criteria are introduced in subsequent chapters if required by the example being considered.

Steel

The steel used in composite bridges tends to be of two primary forms: structural steel in the form of rolled sections or fabricated plates, and bar reinforcement within the concrete element. Occasionally prestressing steel in the form of high-tensile strand or bars may be used. The key properties for design are again the material strengths; these are outlined in Table 1.2.

The limiting stress in tension or compression is similar for small or compact sections but may be less for thicker fabricated sections (Fig. 1.4(a)). A stress–strain curve used for the design of steel elements is shown in Fig. 1.4(b).

The steel elastic modulus is required in composite structures to determine the distribution of load between each material. The modulus for steel (E_a) is generally constant for all steel grades and can be taken as $210\,kN/mm^2$ for plates; however, there may be some slight variation with temperature. For bar reinforcement the modulus is slightly lower at $200\,kN/mm^2$.

Structural steel fabrications are prone to buckling, and the buckling of steelwork elements when compressed may occur in a number of ways. Local buckling of the flange or stiffener outstand is suppressed by the use of outstand ratios. Table 1.3 summarises the key requirements for outstand ratios to prevent local buckling. Where flanges or webs are larger than these limits stiffeners can be applied to suppress the buckling tendency. For composite flanges, connectors may suppress local buckling, provided minimum spacing requirements are met (Table 1.3).

The entire steel section may be prone to buckling under compressive loads and is normally braced to limit this buckling tendency. For columns and members loaded primarily in compression the classic Euler buckling is assumed and bracing provided to limit the effective length of the member. For beams, the buckling may be a lateral torsional form [35], where the instability of the compression

Table 1.2 Strengths for various steel components

Component	Yield stress: N/mm² (or ultimate tensile stress where noted *)
Universal beams, channels and angles. Grade S275	275
Universal beams, channels and angles. Grade S355	355
Plates and flats. Grade S355	275–355, see Fig. 1.4
Mild steel reinforcing bar (plain)	250
High yield reinforcing bar (ribbed)	460
Prestressing bars	880–1200*
HSFG bolts	880*
Prestressing strand	1770–1880*

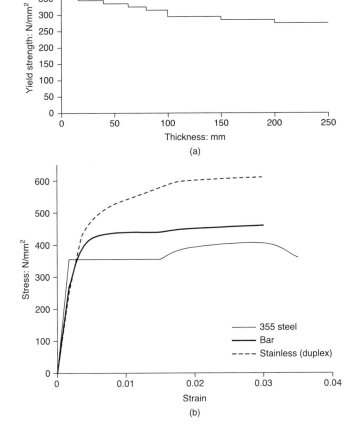

Figure 1.4 (a) Yield strength for grade S355 steel plate of various thicknesses; (b) steel stress–strain curve for various elements.

Table 1.3 *Geometric limits to prevent local buckling of steel plates [2, 4]*

Element	Limit
Slender steel flange in compression. Compact steel	$12t_f$
flange in compression	$7t_f$
Steel flange in tension	$16t_f$
Slender steel web in bending	$228t_w$
Compact steel web in compression	$30t_w$
Compact steel flange of small box girder	$30t_f$
Plate stiffener	$10t_s$
Steel circular section (Non-compact)	$D_0 = 60t_o$
(Compact)	$D_0 = 46t_o$
Connector spacing to non-compact element	4 times connector height or 600 mm
Connector spacing to compact element	$15t_f$ longitudinal
	$30t_f$ transverse
Composite concrete filled box (without connectors)	$42t_w$
Composite circular section (without connectors)	$D_0 = 70t_o$

7

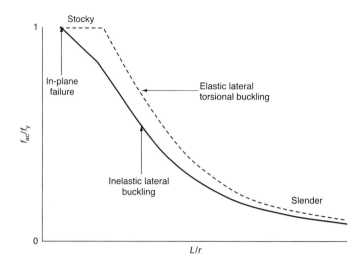

Figure 1.5 Limiting compressive stress for buckling.

flange leads to lateral movement of the whole section. The tendency to buckle can be estimated from the slenderness parameter λ, which is a function of the ratios of applied force (or moment) to the critical buckling force, this force being dependent on the effective length and section properties, primarily the radius of gyration. Figure 1.5 shows the typical reductions in stresses required to limit the tendency to buckle.

Typical effective lengths for elements of composite bridge design are outlined in Table 1.4. The buckling of the steel section in many composite structures is most likely to occur during construction, when the concrete loads the steelwork but the concrete has not hardened and so provides no restraint.

For many steel–concrete composite structures the steel forms a major element of the structure, the design issues most commonly encountered are bending, shear and axial loads. For a steel element subject to a tensile load the maximum stress in the

Table 1.4 Typical effective lengths for composite structures

Element	Effective length	
Steel girders, prior to casting slab	$1.0L_B$	
Compression chord of truss Truss diagonal	$0.85L_C$ $0.7L_D$	
Column fixed to foundation but free at top	$2.0L_C$	
Column fixed to foundation but pinned at top	$1.5L_C$	
Column fixed to foundation and integral with deck	$1.0L_C$	

steel is limited to $0.95f_y$:

$$N_D = 0.95f_y A_{ae} \tag{1.7a}$$

where A_{ae} is the effective area of the steel allowing for any bolt holes.

For a compressive load the maximum stress will be limited to $0.95f_{ac}$, where f_{ac} will be determined from the slenderness parameter and limiting compressive stress curves similar to Fig. 1.5.

$$N_D = 0.95f_{ac} A_a \tag{1.7b}$$

For a steel beam subject to bending, the moment of resistance (M_u) is given by:

$$M_D = 0.95f_y Z \text{ or } 0.95f_{ac} Z \tag{1.8}$$

where Z is the section modulus. Compact sections can be designed assuming fully plastic section properties (Z_p), semi-compact sections use the full elastic section properties, and slender sections need to take into account the effective reductions in web or flange areas that may occur due to out-of-plane buckling and shear lag. Section moduli for standard rolled sections are pre-calculated [24]; for fabricated sections the properties will need to be calculated by the designer. Appendix B outlines the calculation of elastic section properties and Appendix D the calculation of plastic section properties.

For a steel beam with a depth to thickness (d/t) ratio of less than 55 the ultimate design shear resistance of the section is:

$$V_y = 0.95v_y dt \tag{1.9}$$

where v_y is the limiting shear stress. For webs with a slenderness (d/t) ratio of less than 55 the full shear capacity can be assumed:

$$v_y = \frac{f_y}{\sqrt{3}} \tag{1.10}$$

For web depth to thickness (d/t) ratios greater than 55 there may be a reduction in capacity due to web buckling effects. Figure 1.6 shows a typical web shear capacity curve, the limiting values depending on the slenderness (d/t), the web panel ratio (a/d) and flange stiffness (m).

Most steel sections used in composite structures are fabricated sections. The fabrication of a girder or steelwork element involves the assembly of the pieces of steel that will form it in a factory or workshop. The process starts with taking the design and breaking it down into its component elements. For example the simple girder shown in Fig. 1.7 is made up of 12 elements: a top and bottom flange, a web, two end plates, four bearing stiffener plates, two intermediate

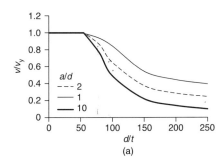

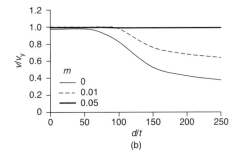

Figure 1.6 Typical limiting shear capacity curves for steel beams: (a) variation with panel geometry; (b) variation with flange stiffness.

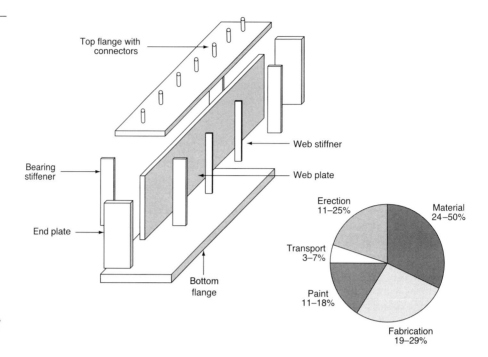

Figure 1.7 Typical elements making up a steel girder.

stiffeners and the shear connectors. A drawing of each element will be produced thus enabling the cutting area of the fabrication shop to produce each part. In many factories the process has been automated and the dimensions of the part are transferred to a cutting machine without the need for a paper copy of the drawing.

Once cut the plates are assembled. The main girder is likely to be put together in one of two ways, using a T&I machine or by using the side-to-side method. The T&I machine takes a web and flange and welds them together to form an inverted T section. This is turned, and the other flange added and welded to form an I section. The side-to-side method takes all three elements. These are clamped together, and with the girder on its side the welds on that side are then laid. The beam is then turned to the other side and the welds there laid. In practice the process is more complex as plates come in 6 m to 18 m lengths and so each flange or web will be formed from a number of plates. The structure is also likely to be larger than can be assembled in the factory or transported easily and so it is broken down into a number of sub-parts.

Nowadays, plates are joined primarily by welding, which involves the laying of molten metal along joints. When cooled this metal has fused with the plates on each side to form a joint. There are a number of processes in which the weld can be formed, submerged arc welding (SAW), metal active gas (MAG) welding and metal arc welding (MAW). The MAW and SAW processes are used for the more automated welding, and the MAG process is the most widely used manual welding method. For the bridge designer there are usually two weld types, a butt weld and a fillet weld. The welds are formulated to have similar properties to the parent metal being joined, such that limiting yield and shear stresses are unaffected. The welding and fabrication process is complex and significant testing and quality control regimes should be in place [25]. On drawings and sketches produced by the designer, welding is normally indicated by shorthand symbols.

Fatigue is a phenomenon primarily affecting the steel elements of a composite structure. Fatigue is primarily influenced by the stress fluctuations in an element (the maximum range of stress, as opposed to the maximum stress), the structure geometry and the number of load cycles. The stress range prior to significant damage varies with the number of load cycles. At high stress ranges the number of cycles to damage is low; there is a stress range below which an indefinitely large number of cycles can be sustained. For design it is critically important to obtain a detail with the maximum fatigue resistance. This is achieved by avoiding sudden changes in stiffness or section thickness, partial penetration welds, intermittent welding or localised attachments.

Composite highway structures are not particularly sensitive to fatigue problems if properly detailed. The design of shear connectors near midspan may be governed by fatigue where static strengths dictate only minimum requirements. Railway bridges with their higher ratio of live to permanent loads are more sensitive to fatigue.

At low temperatures steel can become brittle. Thicker sections containing more impurities or laminations are more likely to have significant residual stresses during fabrication and are the most likely to be affected. Charpy impact testing values are used to measure this; for bridge works in the UK a minimum Charpy value of 27 joules is required. For temperatures below about $-20\,°C$ or sections above 65 mm thickness subjected to a tensile stress a grade of steel with higher Charpy values may be required.

Composite action

There are two primary points to consider when looking at the basic behaviour of a composite structure:

* The differences between the materials.
* The connection of the two materials.

The modular ratio

Differences between the strength and stiffness of the materials acting compositely affect the distribution of load in the structure. Stronger, stiffer materials such as steel attract proportionally more load than does concrete. In order to take such differences into account it is common practice to transform the properties of one material into those of another by the use of the modular ratio.

At working or serviceability loads the structure is likely to be within the elastic limit and the modular ratio is the ratio of the elastic modulus of the materials. For a steel–concrete composite the modular ratio is:

$$n = \frac{E_a}{E_c} \quad \text{or} \quad \frac{E_a}{E_{c'}} \tag{1.11}$$

The value of this ratio varies from 6 to 18 depending on whether the short-term or long-term creep-affected properties of concrete are utilised. Typical values of concrete modulus are given in Table 1.1. It should also be noted that the E_c quoted is normally the instantaneous value at low strain; lower values based on the secant modulus may be appropriate at higher strains. At ultimate loads the modular ratio is the ratio of material strengths; this ratio is dependent on the grade of steel and concrete utilised. For design the different material factors will need to be considered to ensure a safe structure. For a steel–concrete composite:

$$n = \frac{R_a}{R_c} \quad (1.12)$$

where R_a and R_c are the ultimate strengths of steel and concrete. These values will depend on the codes of practice being used and the value of the partial factors, but the modular ratio is approximately

$$n = 2.35 \frac{f_y}{f_{cu}} \quad (1.13)$$

Interface connection

The connection of the two parts of the composite structure is of vital importance; if there is no connection then the two parts will behave independently. If adequately connected, the two parts act as one whole structure, potentially greatly increasing the structure's efficiency.

Imagine a small bridge consisting of two timber planks placed one on another, spanning a small stream. If the interface between the two were smooth and no connecting devices were provided the planks would act independently, there would be significant movement at the interface and each plank would for all practical purposes carry its own weight and half of the imposed loads. If the planks were subsequently nailed together such that there could be no movement at the interface between them then the two parts would be acting compositely and the structure would have an increased section to resist the loads and would be able to carry about twice the load of the non-composite planks. The deflections on the composite structure would also be smaller by a factor of approximately 4, the composite whole being substantially stronger and stiffer than the sum of the parts. A large part of the criteria in the following chapters is aimed at ensuring this connection between parts is adequate.

The force transfer at the interface for composite sections is related to the rate of change of force in the element above the connection. The longitudinal shear flow Q_l is:

$$Q_l = \frac{dy}{dx} N \, dx \quad (1.14)$$

Considering a simple composite beam at the ultimate limit state, assuming it is carrying its maximum force as in Equation (1.4), the maximum change in force in the slab over a length from the support to midspan is:

$$F = 0.4 f_{cu} bt \quad \text{or} \quad 0.57 f_{ck} bt$$

and the shear flow at this stage is:

$$Q_l = 2 \frac{F}{L} \quad (1.15)$$

If the section is capable of significant plastic deformation the shear flow may be considered to be uniform. A consideration of Equation (1.14) will yield that for a simple beam the rate of change of the force in the slab is proportional to the rate of change of moment, or the shear force. For most sections the number of connectors should generally follow the shape of the shear diagram; typically codes allow a 10–20% variation from the elastic shear distribution

$$Q_l = \frac{V A_c y}{I} \quad (1.16)$$

Shear connectors

Shear connectors are devices for ensuring force transfer at the steel–concrete interface; they carry the shear and any coexistent tension between the materials. Without connectors, slip would occur at low stresses. Connectors are of two basic forms, flexible or rigid. Flexible connectors such as headed studs (Fig. 1.8) behave in a ductile manner allowing significant movement or slip at the ultimate limit state. The number of connectors required (No) is determined by dividing the longitudinal shear force by the capacity of a connector, at the ultimate limit state:

$$No = \frac{Q_1}{0.71 P_u} \tag{1.17a}$$

At the serviceability limit state the loads on the connectors should be limited to approximately half the connector's static strength to limit slip.

$$No = \frac{Q_1}{0.55 P_u} \tag{1.17b}$$

The type of connector used should reflect the type of load. For normal relatively uniform shear flows, stud connectors are economic. For heavier shear flows, bar or perforated plate connectors may be more applicable. Where shear flows are more concentrated with sudden changes in magnitude, the larger rigid plate connectors are more appropriate. Where there is the likelihood of tensile loads occurring, shear hoop connectors or long studs should be used such that the tensile load can be resisted in the main body of concrete and suitable reinforcement detailed around the hoop or stud head. Where shear connectors are used with coexistent tension, the shear capacity should be reduced:

$$Q_{1\,max} = (Q_1 + 0.33T^2)^{0.5} \tag{1.18}$$

Where the tension exceeds about 10% of the shear, detailing of the tensile resistance load path should be considered with the use of longer connectors or additional tensile link reinforcement in the concrete.

Rigid connectors such as fabricated steel blocks or bars behave in a more brittle fashion; failure is either by fracture of the weld connecting the device to the beam or by local crushing of the concrete. Typical nominal static strengths P_u for various connector types for grade $32/40\,\text{N/mm}^2$ concrete are given in Table 1.5. Other

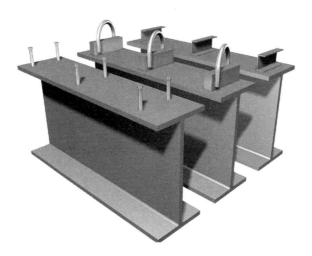

Figure 1.8 Typical shear connector types for steel–concrete composite construction, studs, bars with hoops and channels.

13

Table 1.5 *Nominal static strengths of shear connectors*

Type of connector	Connector material	Nominal static strength per connector for concrete grade 32/40
Headed studs, 100 mm or more in height, and diameter:	Steel $f_y = 385$ N/mm^2 and minimum elongation of 18%	
19 mm		109 kN
22 mm		139 kN
25 mm		168 kN
50 × 40 × 200 mm bar with hoops	Steel $f_y = 250$ N/mm^2	963 kN
Channels:	Steel $f_y = 250$ N/mm^2	
127 × 64 × 14.9 kg × 150 mm		419 kN
102 × 51 × 10.4 kg × 150 mm		364 kN

forms of shear connector are also available, for example perforated plates [17, 26], undulating plates or toothed plates [27].

The capacity of the connectors depends upon a number of variables including the material strength, the stiffness (of the connector, steel girder and concrete), the width, spacing and height of the connector. For the common forms of connectors outlined in Fig. 1.8 and Table 1.5 the nominal static strengths can be estimated from the following equations.

Shear studs:

$$P_u = 11f_{ck}d^2 \tag{1.19a}$$

$$P_u = 0.55f_{ua}d^2 \tag{1.19b}$$

For shear studs with an ultimate strength f_{ua}, of diameter d and a height of more than $4d$, the lesser of Equations (1.19a) and (1.19b) should be used.

Channel connector:

$$P_u = 0.7A_f f_{ck} \tag{1.20}$$

for the relatively flexible channel connectors, where A_f is the projected area (Bh).

For the stiffer bar connectors:

$$P_u = 2.5A_f f_{ck} \tag{1.21}$$

For hoop connectors:

$$P_u = 1.1f_y d^2 \tag{1.22}$$

For the combination of hoop and bar connectors of Table 1.5, the capacity is that of the bar plus 70% of the hoop. The capacity of the bar or channel connector may also be governed by the size of weld between the connector and girder:

$$P_u = (1.2D + 1.45B)f_y w \tag{1.23}$$

where B is the width, D is the depth (along the girder) of the connector, w is the weld size and f_y is the yield strength of the weld (taken as the lower of the girder or connector strength).

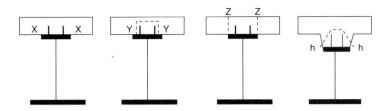

Figure 1.9 Typical shear planes in a steel–concrete composite structure.

At the ultimate limit state the failure of shear planes (see Fig. 1.9) other than at the interface (x–x) may need to be investigated. These shear planes will be around the connector (y–y), or through the slab (z–z). Where haunches are used a check on shear planes (h–h) through the haunch may be required.

The shear planes around the connectors rely on the shear strength of concrete and any reinforcement passing through the shear planes.

$$Q_1 = 0.9L_s + 0.7A_s f_y \qquad (1.24)$$

where L_s is the length of the relevant shear plane. Typically, $0.17\% A_c$ of reinforcement should be provided through the shear plane to ensure a robust structure with a ductile behaviour. The maximum force that can be carried by the shear plane is limited by the concrete strength.

$$Q_1 = 0.15f_{cu}L_s \qquad (1.25a)$$
$$Q_1 = 0.125f_{ck}L_s \qquad (1.25b)$$

Where this limit governs, a wider or taller connector layout is required.

2

Simple beam bridges

...estimation of flange size assumes that the flanges will primarily carry the moments with only a small contribution from the web...

Introduction

Simply supported steel beam and concrete slab bridges are a common form of construction for spans to about 50 m. At the lower range Universal Beam sections may be used; for spans above about 20 m a fabricated girder is likely to be economic. The form of the structure consists of a number of beams placed side by side (usually parallel and in pairs) at 2 m to 4 m spacing. The beams will be linked by bracing to prevent buckling instability during construction. A concrete slab about 200 to 250 mm thick is constructed on the beams. The slab carries the highway (or railway) loading and spreads it to the underlying beams. To ensure composite action between the slab and steelwork, connectors are provided at the steel–concrete interface.

Initial sizing

The initial sizing of the structural elements of a bridge is a mixture of the use of geometric ratios (as outlined in Table 2.1), previous experience (of the designer or his/her peers) and approximate calculations. Charts giving flange- and web sizes for various spans are available; however, they do not convey any understanding of loads or behaviour and should be used with care. The author prefers a more direct calculation method as outlined below; it conveys some understanding of what is likely to be critical in the detailed design that will follow. The method can be modified to allow for various loading types and to include the more complex truss, arch or stayed forms outlined in subsequent chapters.

Example 2.1

In this chapter the preliminary steel, concrete and connector designs are illustrated using the Quakers Yard Bridge as an example. The bridge is shown in Fig. 2.1, it is a simply supported bridge that carries a road over a small river in a Welsh valley. The bridge is designed to accommodate the movement and settlement induced by mining in the underlying bedrock [28], which is achieved using bearings and joints at each end to accommodate the movement.

The structure was designed in the early 1980s and was one of the first structures to be designed to BS 5400 [2, 3, 4] rather than to the earlier codes, BS 153 [29] or CP 117 [30]. The structure was reviewed by senior engineers with experience of

Table 2.1 Typical span–depth ratios for simply supported steel–concrete girder bridges

Element	Span-to-depth ratio	Remarks
Main steel girders for multi-beam layout	18 to 22 10 to 20	Highway bridges Railway bridges
Deck slab spanning between girders	10 to 25 4 to 20	Highway bridges Railway bridges 200 mm typical minimum to achieve cover requirements to reinforcement
Deck slab cantilevers outside girders	5 to 10	Thickness dependent on parapet type, for high-containment parapets 250 mm or more required

BS 153 and so the details of the structure were heavily influenced by the earlier code, particularly the web which is more slender and more regularly stiffened than would be normal nowadays. More compact sections are now common. This, to some degree, illustrates the influence of codes on design. As engineers become more proficient and understand the boundaries of the rules in the codes they are using, then they use the rules to develop and refine forms. At the time of writing this chapter the UK is starting to use Eurocodes, and similar influences are likely over the next few years, leading to further development in structural details or form to suit these rules.

Having a span of 28–30 m and a width of 12–13 m, the depths of structural members can be estimated from span-to-depth ratios (Table 2.1). Girder spacing is typically 3 m to 4 m for highway bridges, and for edge cantilevers 1 m to 1.5 m. An arrangement with four girders at 3.2 m centres and a 1.2 m edge cantilever was chosen based on these rules. For cantilever slabs span-to-depth ratios of 6 to 8 are typical and for internal spans ratios of 12 to 25 are feasible.

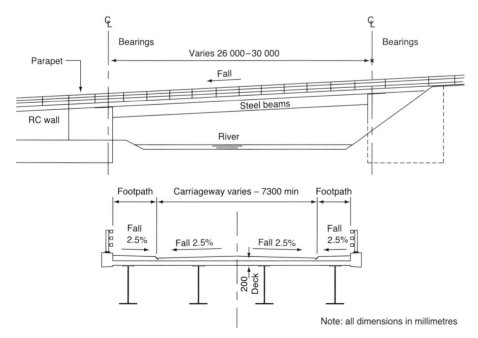

Figure 2.1 Layout of Quakers Yard Bridge.

A slab depth of 200 mm was chosen. It should be noted that deeper slabs require less reinforcement, particularly if they allow arching to occur (see Chapter 5); however, as the slab is a major part of the dead load of the bridge an over-deep slab will add to the overall steelwork tonnage. For the girders a span-to-depth ratio of 16 to 22 is economic.

Loads

Figure 2.2 outlines a typical steelwork tonnage for steel plate girder bridges constructed in the UK. A first estimate of the steelwork tonnage from Fig. 2.3 for a highway bridge gives a range of 130–160 kg/m^2; a figure of 150 kg/m^2 (1.5 kN/m^2) is used for this example. Dead loads are calculated from the slab and surfacing dimensions, assuming densities of 25 kN/m^3 for concrete and 23 kN/m^3 for surfacing. Live loads are obtained from the loading standard [31, 32]; in this case HA and 45 units of HB will be considered.

An HA loading is a uniform load of varying intensity intended to model a carriageway full of vehicles; it includes for overloading of vehicles, impact effects and the bunching of vehicles across lane markings (see also Chapter 9). For this example HA loading is 35 kN/m for each lane. A small knife-edge load of 120 kN/lane is used in association with the uniform load. The HA loads tend to govern the design of longer spans. An HB loading is a single heavy vehicle load, which may occur alone or in combination with HA loads. The HB loading tends to govern the design of smaller spans, or local beams and slab elements of larger structures. The HB loading consists of a four-axle vehicle with loads of 25 to 45 tonnes per axle (25 to 45 units). For a structure on a minor road, the HB capacity would normally be about 30 units. For our bridge if the adjacent main road were closed it would form part of a diversion route, and so the full 45 units are considered appropriate. For bridges of spans 15 to 75 m a combination of HB vehicle with HA loading in the adjacent lanes is likely to be critical. The total permanent (G) and variable (Q) loads are calculated from the structure geometry:

Steel	$G_a = 1.5 \times 3.2 \times 29$	$= 139 \, \text{kN}$
Concrete	$G_c = 3.2 \times 0.2 \times 25 \times 29$	$= 464 \, \text{kN}$
Surfacing	$G_{sdl} = 3.2 \times 0.14 \times 23 \times 29$	$= 299 \, \text{kN}$
Live (HA)	$Q_{ha} = (0.92 \times 35 \times 29) + 120$	$= 1054 \, \text{kN}$
Live (HB)	$Q_{hb} = 4 \times 450$	$= 1800 \, \text{kN}$

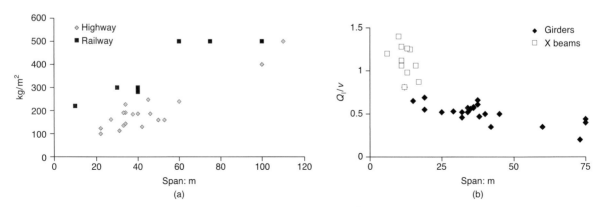

Figure 2.2 (a) Typical steel weight per square metre of deck; (b) typical Q_l/V ratios for various spans.

Table 2.2 Loading on non-composite and composite elements of Example 2.1

Element	Load: kN	ULS factor	ULS load: kN
Steel	139	1.05 × 1.1	160
Concrete	464	1.15 × 1.1	586
Subtotal (on non-composite section)			746
Surfacing	299	1.75 × 1.1	575
HA	527	1.5 × 1.1	869
HB	900	1.3 × 1.1	1287
Subtotal (on composite section)			2731

Analysis

For this simply supported structure the analysis is relatively simple. A grillage analysis [45] could be carried out or a simpler D-type analysis [33] to estimate the load spread across the deck. For this structure it is sufficient to estimate the proportion of load carried by a girder and to analyse a statically determinant simply supported line beam structure. For this analysis the full dead and surfacing loads will be used, the live load on the girder is assumed to comprise 50% of the full HA lane load and 50% of the HB vehicle.

The partial load factors for the loads are obtained from the design standard [31, 32] and the factored loads for analysis calculated in Table 2.2. The moment and shear diagrams for these loads are shown in Fig. 2.3.

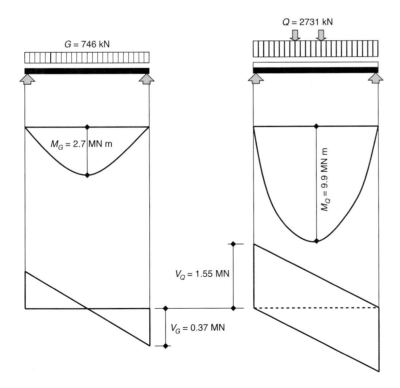

Figure 2.3 Bending and shear diagrams for Example 2.1 at the ultimate limit state.

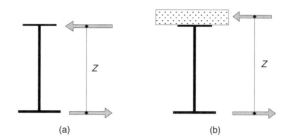

Figure 2.4 Force couples for the beam: (a) non-composite; (b) composite.

Initial design of girder

In this section the estimates of girder flange and web sizes will be made. Consider the idealised behaviour of a simply supported composite bridge comprising a steel girder with a concrete top slab. Initially, on completion of the slab construction, only the unpropped steel section is stressed (ignoring at this stage the effects of shrinkage) and there is no force transfer at the steel–concrete interface. For loads added after this stage, the composite section will carry them. Stresses in the beam increase, stresses in the slab occur and there is a force transfer at the steel–concrete interface. The method of estimation of flange size assumes that the flanges will primarily carry the moments with only a small contribution from the web [34] (see Appendix A):

$$M = z N_{\mathrm{C}} \tag{2.1a}$$

$$M = z N_{\mathrm{T}} \tag{2.1b}$$

where z is the distance between flanges and N_{C} and N_{T} are the compression or tension in the flange (Fig. 2.4).

During construction of the bridge the maximum load occurs as the slab is cast, and the girder is non-composite. From Fig. 2.3, $M = 2.7\,\mathrm{MN\,m}$, and rearranging Equation (2.1):

$$N_{\mathrm{T}} = N_{\mathrm{C}} = \frac{M}{z} = \frac{2.7}{1.6} = 1.7\,\mathrm{MN}$$

Having determined the force in the flange, dividing this by a suitable stress will allow the flange area to be estimated:

$$A = \frac{N}{f} \tag{2.2}$$

At this point we need to consider carefully how the bridge is being built. For steel–concrete composite bridges the methods and sequences of construction are vitally important to the choice of limiting stress.

Bracing of the steelwork

The steelwork is relatively slender and usually requires bracing to ensure stability. The bracing must be designed to ensure all likely buckling modes are suppressed, this includes both instability of the girders between the bracing points, and overall instability of a braced pair or of the whole bridge.

For an individual beam or girder the main mode of instability is lateral torsional buckling [35], where the beam undergoes a simultaneous lateral movement and rotation. The tendency for a beam to buckle is influenced by a number of factors including: the nature of the external loads, the beam shape, the bracing type strength and stiffness, fabrication tolerances, the length between effective bracing, residual stresses (from rolling or welding) and the stress in the compression flange. Bracing in the form of plan bracing (Fig. 2.5(d)), anchoring to a rigid object, transverse torsional bracing (Fig. 2.5(b) and (c)) or a combination of these methods can be used to suppress lateral torsional buckling. Some forms of bracing are outlined in Fig. 2.5, and it should be noted that simply linking two unstable compression flanges (Fig. 2.5(a)) would not be effective.

In order for the bracing to function it must have sufficient strength to resist the applied loads. The destabilising forces generated by the compression flange are generally considered to be approximately 2.5% of the sum of the forces in the flanges being connected, or half of this value when considered with wind or other significant transverse effects. The force can vary significantly, the more flexible the bracing the larger is the destabilising force to be resisted [36].

Typically bracing is required at 12 to 15 times the flange width, with 8 m being a typical minimum and 16 m being a typical maximum. The determination of limiting moments or stresses is based on a slenderness parameter (λ) using an effective length concept, and the limiting compressive stress is determined from Fig. 2.6. Beams with λ of 45 or less can generate the full beam capacity; beams with λ greater than 45 are prone to some instability and stresses must be limited.

Where the overall slenderness of the section is critical, either plan bracing is required or a more stable box section should be used; alternatively adjacent pairs of girders should be connected. When more than two girders are connected, the designer should be aware that the bracing will participate in the transverse spreading of load in the final condition. For this participating bracing the forces

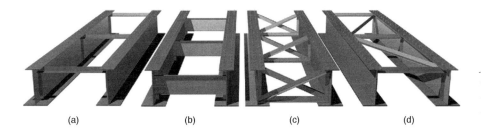

(a)　　　　　(b)　　　　　(c)　　　　　(d)

Figure 2.5 Bracing types for steel–concrete composite construction: (a) ineffective bracing; (b) U-frame bracing; (c) cross-bracing; (d) plan bracing.

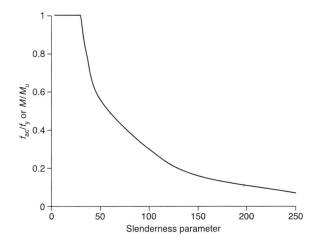

Figure 2.6 Limiting compressive stresses for a steel girder.

in the final condition may govern its design and the detailing of the connections with the slab and main girders. The designer must also consider the effects of fatigue, particularly at the areas adjacent to the web, flange and bracing connections.

In our example we have now to determine the bracing type and flange size. In the permanent condition the span is to be torsionally flexible such that it can accommodate mining movement; no bracing will be left in permanently except at the abutments where a flexible U-frame [37] restraint will be used. A system of plan bracing is chosen for the temporary condition, as this will also improve the overall stability of the girder pair. To limit the angle of the plan braces (such that it is not too acute) the length between horizontals will be about 7 m (see Fig. 2.7).

Since the structure is well braced we will initially assume a conservative limiting compressive stress of about half of the maximum for a grade 355 steel, 170 N/mm². Our compressive force was 1.7 MN so the top flange area A_{at} (using Equation (2.2)) is:

$$A_{at} = \frac{N_C}{f_{ac}} = \frac{1\,700\,000}{170} = 10\,000 \, \text{mm}^2$$

The minimum outstand is $12t_f$ (Table 1.3), giving a width of $24t_f$, so the optimum thickness is:

$$t_f = \left(\frac{A_{at}}{24} \right)^{0.5} = 21 \, \text{mm}$$

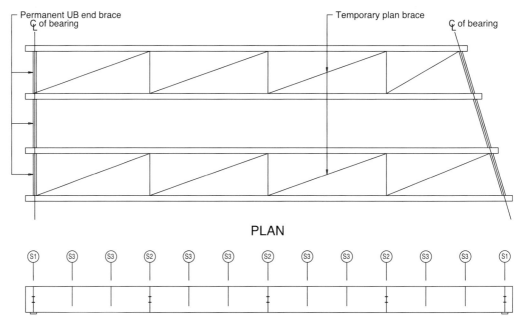

Shear connectors	3 @ 150	3 @ 300	3 @ 150
Top flange		500×20	
Web plate		1600×12	
Bottom flange		800×25	

Figure 2.7 Bracing layout for Example 2.1.

So a 500 × 20 mm top flange should be satisfactory when allowing some contribution from the web (see Appendix A).

The surfacing and live loads are carried by the completed composite structure. Again the moments are carried primarily by the flanges; from Fig. 2.3, $M = 9.9\,\text{MN}\,\text{m}$.

$$N_T = \frac{M}{Z} = \frac{9.9}{1.7} = 5.8\,\text{MN}$$

The bottom flange will carry the tensile forces from both the composite and non-composite conditions, $N_T = 1.7 + 5.8 = 7.5\,\text{MN}$.

The limiting tensile stress is $0.95 f_y = 0.95 \times 345 = 327\,\text{N}/\text{mm}^2$ for a grade 355 steel of thickness less than 40 mm (Fig. 1.4(a)). Using Equation (2.2) again:

$$A_{at} = \frac{7\,500\,000}{327} = 22\,900\,\text{mm}^2$$

Allowing that the web will carry a small proportion of this force a flange area of $20\,000\,\text{mm}^2$ is used. For overall stability a wide flange is preferred, for tension flanges the maximum outstand is $16 t_f$, giving a width of $32 t_f$ and so the optimum thickness is:

$$t_f = \left(\frac{A_{at}}{32}\right)^{0.5} = 25\,\text{mm}$$

So an 800 × 25 mm bottom flange should be satisfactory.

Having determined the flange sizes a review of the assumed limiting compressive stress should be made. The limiting stress is taken from Fig. 2.6 and depends upon the beam's slenderness parameter λ.

$$\lambda = \frac{k L_R}{r} \tag{2.3}$$

where L_R is the distance between restraints, r is the girder's radius of gyration and k is a coefficient dependent upon the stiffness of the intermediate restraints, the variation in moment between restraints and the geometry of the beam. For a moderate span simply supported plate girder with the top flange smaller than the bottom flange and with multiple restraints, k may be initially taken to be 1.3 (see Equations (4.4) and (5.4)). The radius of gyration r may be calculated from the beam geometry (see Appendix B) or where top and bottom flange thicknesses are similar this may be taken as:

$$r = 0.25 B' \tag{2.4}$$

where B' is the average flange width (see also Appendix A). For the example, $L_R = 7\,\text{m}$, $B' = 650\,\text{mm}$ and $r = 163\,\text{mm}$. Using Equation (2.3):

$$\lambda = \frac{k L_R}{r} = \frac{1.3 \times 7}{0.163} = 56$$

From Fig. 2.6 the limiting moment or stress is 60% of the yield value; this is greater than the 50% originally assumed and so is satisfactory.

The web of the girder carries the entire shear in the composite and non-composite stages; from Fig. 2.3, $V_{\text{max}} = 1.92\,\text{MN}$. As noted above we are designing using a slender web with stiffeners giving a panel aspect ratio of about 1.5. An initial limiting stress of $100\,\text{N/mm}^2$, half of the maximum, is used as an initial estimate. Rearranging Equation (1.9), the web thickness is:

$$t_{\text{w}} = \frac{V_{\text{max}}}{100d} = \frac{1\,920\,000}{100 \times 1600} = 12\,\text{mm}$$

Checking the limiting stress using this size: $d/t = 133$, $a/d = 1.5$ and from Fig. 1.6, $v/v_{\text{y}} = 0.57$ or $111\,\text{N/mm}^2$. This is higher than the limiting stress used, so a 12 mm web should be satisfactory. If less stiffening were used the web would be thicker, say 15 mm. A thicker web with less stiffeners may be preferred by some UK fabricators.

The web panels are divided by vertical stiffeners, termed transverse stiffeners, as they are usually located transversely to the main web axis. The stiffeners are required to be sufficiently stiff such that all web buckling occurs within the web panel and does not involve significant deformation of the stiffener. The stiffener design is based on strength; studies [38] have shown that this strength-based design provides more than sufficient stiffness. The stiffener should be designed to resist direct loads, loads from tension field action or from a destabilising load from the web. For normal bridge loads the direct load to a stiffener is small. For the design method in the example no significant tension field action is assumed, and so the stiffener need only be sized for destabilising loads:

$$N_{\text{des}} = \frac{L_{\text{S}}^2 k}{a} t_{\text{w}} f_{\text{R}} \tag{2.5}$$

where L_{S} is the stiffener length, t_{w} the web thickness, k is a coefficient dependent on the stiffener slenderness (but with a maximum value of 0.4) and f_{R} is an estimate of the web stress:

$$f_{\text{R}} = v + \frac{f}{6} \tag{2.6}$$

To minimise waste it is normal to assume initially that transverse stiffeners will be formed from plate of the same thickness as the web, which allows them to be cut from major off-cuts from the fabrication process. Using a stiffener of a similar thickness as the web will generally be satisfactory. Figure 2.8 shows the destabilising force for various girder depths. It can be seen that the destabilising force only approaches the stiffener capacity for a girder depth over 3 m (d/t greater than 250). A simple single-leg stiffener is to be used for the design example. The stiffener can have a maximum outstand ratio of 10 (Table 1.3). The stiffener acts with a $32t$ length of web, giving the section shown in Fig. 2.8(b).

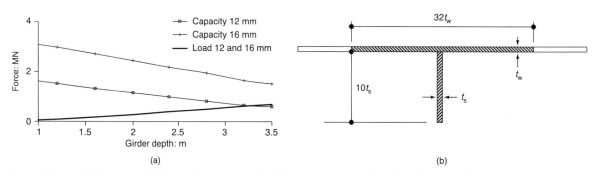

Figure 2.8 (a) Stiffener destabilising load and capacity for 12 mm and 16 mm webs; (b) stiffener details for Example 2.1.

Initial design of the concrete slab

The design of the slab for the deck in transverse bending will be governed by the heavy axles of the HB or abnormal vehicles. For the preliminary design it is sufficient to estimate moments based on a strip method [39]. Allowing for the spread of the wheels, and of the load through the surfacing and slab to the supports, a 4.5 m width (B) is chosen (see also Example 5.1) as shown in Fig. 2.9.

The heavy vehicle will tend to deflect the steel beams directly below it more than those further away, so there is a tendency for the slab to have more negative (sagging) moments. A conservative assumption for the slab design is to assume only small hogging effects over the beam such that the nominal strip can be considered as effectively simply supported. From Fig. 2.9 there are six 112.5 kN wheel loads on the slab strip, with a partial load factor of $1.3 \times 1.1 = 1.43$, then the ultimate load is:

$$Q = 6 \times 112.5 \times 1.43 = 965\,\text{kN}$$

The slab moment is:

$$M = \frac{QL}{8B} = \frac{965 \times 3.2}{8 \times 4.5} = 85\,\text{kN}\,\text{m/m}$$

Using Equation (1.5), with $d = 0.15$ m, the limiting slab moment is:

$$M_{\text{u}} = 0.15 bd^2 f_{\text{cu}} = 0.15 \times 1.0 \times 0.15^2 \times 40 = 0.135\,\text{MNm/m or } 135\,\text{kN}\,\text{m/m}$$

This is greater than the applied moment. The reinforcement required in the slab is estimated from Equation (1.6), rearranging this for A_S and assuming $f_y = 460\,\text{N/mm}^2$ (Table 1.2):

$$A_S = \frac{M}{0.65 df_y} = \frac{0.085}{(0.65 \times 0.15 \times 460)} \quad \text{which gives } 1900\,\text{mm}^2/\text{m}$$

This area is provided by T20 bars at 150 mm centres in the bottom of the slab.

Longitudinally the concrete slab will also have to carry the compression induced by overall bending when the section becomes composite. For the concrete, the maximum force that can be carried is as Equation (1.4), with $f_{\text{cu}} = 40\,\text{N/mm}^2$:

$$N_{C\,\text{max}} = 0.4 bdf_{\text{cu}} = 3.2 \times 0.2 \times 0.4 \times 40 = 10.2\,\text{MN}$$

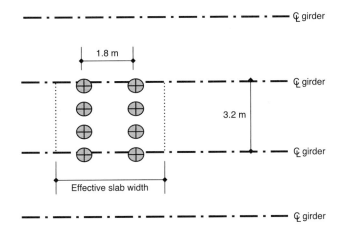

Figure 2.9 Layout of HB vehicle on slab strip.

this is greater than the applied force of 5.8 MN and so the slab is also satisfactory for longitudinal compression loads.

Initial shear connector design

Having designed both the steel and concrete elements, the shear connector numbers need to be estimated. From the previous calculations it was noted that the ultimate force that could be carried by the slab was 10.2 MN, from Equation (1.15) the ultimate shear flow is:

$$Q_l = \frac{2C_{max}}{L} = 2 \times \frac{10.2}{29} = 0.7\,\text{MN/m or } 700\,\text{kN/m}$$

Assuming 19 mm diameter shear studs with a resistance of 109 kN each (see Table 1.5) and using Equation (1.17(a)):

$$No = \frac{Q_l}{0.71 P_u} = \frac{700}{109 \times 0.65} = 10 \text{ per metre}$$

so three studs at 300 mm centres along the beam should be satisfactory. This will give a connector requirement that ensures the connection interface has sufficient strength and is not the weakest link at the ultimate limit state; it also complies with the minimum connector spacing of Table 1.3.

Near the support the connectors will be governed by serviceability limits and the number of connectors required will follow the shape of the shear diagram (Fig. 2.3) and be derived from Equation (1.16). For girders of the size used for this bridge, Ay/I is typically 0.5 to 0.7 (Fig. 2.2(b)), but can be calculated from the flange and web sizes derived (see Appendix C). Using $Q_l = 0.6V$, and assuming that the shear at serviceability is about 75% of the ultimate value:

$$Q_l = 0.6 \times 0.75 \times 1.92 = 0.86\,\text{MN/m or } 860\,\text{kN/m}$$

To prevent slip or undue deformation the shear stud load is limited to 55% of its characteristic strength (Equation (1.17b)):

$$No = \frac{Q_l}{0.55 P_u} = \frac{860}{108 \times 0.55} = 15 \text{ per metre}$$

so three studs at 150 mm centres near the edge of the beam should be satisfactory. This is a slight over-provision but it keeps the connector spacing at a multiple of 150 mm, the same as that derived for the slab reinforcement; this makes detailing and placing of the reinforcement simpler. The detailing requirements for the reinforcement and connectors are considered in more detail in Chapter 3 (Example 3.1).

Safety through design

Throughout the process of design a number of decisions are made that affect the form and safety of the structure. The decisions should be clear in correspondence or reports such that they can be developed at later stages of design and into the construction as part of a safety file. Table 2.3 outlines the hazards, the risk and the decisions associated with them to reduce or mitigate the hazards for this structure.

Table 2.3 *Mitigation of risks for various hazards associated with the initial design of Example 2.1*

Hazard	Risk/severity	Mitigation
Spanning of a fast-flowing river subject to large flow variation		
(a) washing away of falsework or props	Possible with severe consequences	Use prefabricated elements requiring no falsework or propping in the river
(b) flood rising to level of beams	Unlikely but with severe consequences	Define flood levels, provide robustness in the design of the bridge lateral load capacity
Operators falling during construction	Possible with injury	Avoid joints in main steelwork to limit need for access
Inspectors falling during maintenance or repair	Possible with injury	Discuss requirements with maintenance authority
Operators falling during painting	Possible with injury	Consider the use of weathering steel or specify primary painting off site with long-lasting system to minimise repainting requirements
Users falling from the bridge	Possible with injury	Provide parapets to edges with an appropriate containment capacity
Steelwork buckling during construction	Unlikely but with severe consequences	Design bracing as part of structure design, specify construction sequence of steel and concrete deck slab
Movement of ground due to mineral extraction, causing dislocation and falling of span	Unlikely but with severe consequences	Define possible ground movements; design a flexible structure to accommodate significant movement, use bearings and joints with robust movement capacity
Overloading of bridge	Possible if adjacent main road closed	Design the structure for full HB or abnormal vehicles (discuss with maintenance authority)

Environmental issues

When designing a bridge there may be a number of environmental issues, from the impact of the structure on plants and wildlife, landscape and archaeology, or changes to water runoff, river flows and the pollution of waterways. However, a more global environmental concern recently raised is that of the embodied energy use and carbon dioxide (CO_2) emissions of the construction materials. Steel–concrete composite bridges use the two primary construction materials that have relatively high embodied energy in their production and transportation. One of the key environmental objectives should be to minimise the amount of steel and concrete used in any structure. This the engineer can achieve through design, and low volumes of steel and concrete also usually lead to an economic design. Studies indicate that there is little difference in overall embodied energy between steel and concrete for building structures [13].

There are significant advantages if recycled steel is used (it has a primary energy input that is only 65% that of steel derived from iron ore). The use of cement derived from a dry production process similarly has lower energy and CO_2 emissions than cement from a wet production process. Concrete made with some cement substitutes such as pulverised fuel ash or ground granulated blast furnace products can also reduce embodied energy. The use of materials from a local source will also minimise transportation energy and CO_2. All of the above may be controlled by the designer, to some degree, in the specification for materials.

3

Integral bridges

...integral bridges are used for all bridges up to 60 m long...

Introduction

Integral bridges are structures where joints are not required as the substructure and superstructure are monolithic. Integral bridges have significant maintenance advantages over conventionally articulated bridges. Leaking joints are a major source of water ingress [40], causing deterioration. In the UK, the Highways Agency recommends that integral bridges be used for all bridges up to 60 m long unless there are good reasons not to do so [41]. Integral bridges can be used on longer lengths; in the USA, lengths of double this are common in some states.

Integral bridges come in two forms: the semi-integral or fully integral. The semi-integral form has no joints but has bearings to give some rotational flexibility to the end of the deck. The fully integral bridge has no joints or bearings. Ideally the abutment is kept as shallow as possible by the use of a bank-seat-type structure (Fig. 3.1(a)–(c)) limiting the amount of soil affected by the movement. When a bank seat is not viable a full height wall of concrete or steel sheet piles is required, forming a portal structure (Fig. 3.1(d)–(e)).

Bridges move primarily due to thermal effects although concrete shrinkage and vehicle traction may also cause some movement. A bridge will undergo a series of cyclic movements as temperatures fluctuate between day and night and between

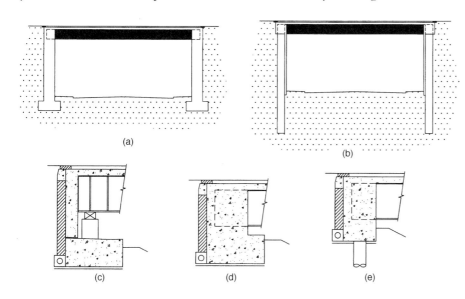

(a)

(b)

(c)   (d)   (e)

Figure 3.1 Portal structure: (a) on pad foundations; (b) on embedded piles; (c) bank-seat abutment, semi-integral; (d) integral; (e) piled integral.

summer and winter. The thermal mass of a structure has some influence in damping the fluctuations. In the UK, typical thermal strains are ±0.0004 for concrete bridges and ±0.0006 for steel bridges with steel–concrete composite bridges being between the two [41]. More detailed thermal strains can be estimated using the coefficient of thermal expansion and the likely extreme bridge temperatures [32]. For structures located in the south of England or which use concrete with a lower coefficient of thermal expansion (for instance limestone or lightweight aggregates) some reduction in the thermal strains may be found.

Soil–structure interaction

If the likely thermal movement is in the range of ±20 mm an asphaltic plug [43] in the carriageway surfacing will be capable of accommodating any tendency to crack and no expansion joint is required. With no expansion joint a bridge structure will be in contact with the soil backfill, and bridge movement will induce pressures in the soil. These pressures will in turn need to be resisted by the structure. It is common to visualise the soil–structure interaction as a change in soil pressures between active (K_a) and passive (K_p), with the static pressure being near the at-rest (K_o) state.

For vertical walls with a drainage layer behind the wall, such that there is little friction between wall and soil, the soil pressure is assumed to act perpendicular to the wall. The various soil pressure coefficients can be estimated from the following equations:

$$K_a = \frac{1 - \sin\phi}{1 + \sin\phi} \tag{3.1a}$$

$$K_o = 1 - \tan\phi \tag{3.1b}$$

$$K_p = \frac{1 + \sin\phi}{1 - \sin\phi} \tag{3.1c}$$

where ϕ is the soil angle of shear resistance. If there is some friction between wall and soil there will be some change to the soil pressure coefficients, slightly reducing the active but significantly increasing the passive. A friction angle of 50 to 75% of the soil shear strength is generally assumed; values of K_p [41] taking wall friction into account are shown in Fig. 3.2(a).

The pressures induced by the thermal movements depend primarily upon the level of movement of the bridge and the properties of the soil. Traditionally,

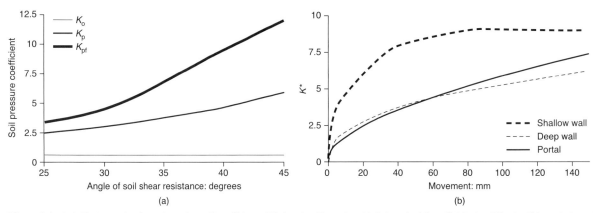

Figure 3.2 (a) Estimated values of passive soil coefficients (K_p) using Equation (3.1c) and with wall friction (K_{pf}); (b) variation of soil pressure with movement for two wall heights.

backfill behind bridge abutments has been of a good quality, usually better than the adjacent embankments. Fill of good quality will, however, generate very high passive soil pressures, leading to larger forces on the structure. For integral bridges it is normal to specify a maximum soil strength to avoid very high passive pressures. Care needs to be taken in setting the minimum soil strength as the fill will need to resist the highway loads and too weak a fill will not give a good transition between the embankment and structure.

The backfill behaves in a non-linear manner during loading; on repeated loading it will tend to undergo some volume change and residual strains will occur, with the accumulation of strain occurring in a ratcheting effect [44]. The relationship between maximum soil pressure (K^*) and the wall movement (Δ) is normally expressed as:

$$K^* = K_o + \text{function}(K_p, \Delta) \tag{3.2}$$

The function of K_p depends on the wall height and form. Figure 3.2(b) shows a typical variation of K^* for a wall with a $K_p = 9$ and wall heights of 2.5 m and 7.5 m.

Example 3.1

The first example in this chapter is another simply supported river bridge, this time of a semi-integral form. The bridge was designed in the early 2000s as part of a new motorway around Birmingham. The bridge is shown in Fig. 3.3. It is a

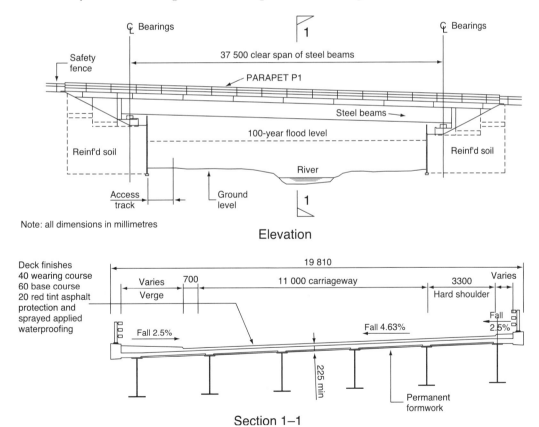

Figure 3.3 River Blythe Bridge details.

multi-beam form of 37.5 m span, which is formed from weathering steel girders. The bridge abutments are formed from reinforced earth walls with a bank seat supporting the bridge deck. The bridge is semi-integral rather than fully integral to limit the additional strains on the reinforced earth substructure. The bridge has been sized (in a similar way to the example of Chapter 2) for HA loading and 30 units of HB.

Loads

The dead loads of the steelwork, concrete and surfacing can be determined from the outline sizing and the layout drawing. Weights of parapets and parapet coping together with additional surfacing at the footway will also be established for this detailed stage. Live loads for HA, HB and the footway load are determined from the codes [32]. The soil loading can be established from Fig. 3.2 or by using the equations in the codes [41]. To estimate the soil loading, the temperature movement and soil characteristics must be known. The bridge bank seat is supported on a re-inforced soil wall with a selected granular material. The soil fill behind the bank seat is of a similar material with a density of 19 kN/m^3 and angle of shearing resistance of 42°, giving an estimated K_p of between 5 and 9 (Fig. 3.2(a)). Assuming a thermal strain of 0.0005 for a steel–concrete composite structure then the movement for a 37.5 m long structure is 10 mm (0.005 × 37 500/2). A conservative estimate of the maximum soil pressure coefficient K^* is 4.5 from Fig. 3.2(b). The peak soil pressure at the base of the abutment wall is:

$$q = K^* \rho H \tag{3.3}$$

The wall height (H) is 2.5 m and the soil density (ρ) is 19 kN/m^2, so for this example:

$$q = K^* \rho H = 4.5 \times 19 \times 2.5 = 214 \, \text{kN/m}^2$$

The maximum soil force, assuming a triangular pressure distribution is:

$$Q_s = 0.5qH \tag{3.4}$$

which for this example gives a characteristic force of 267 kN/m due to soil loading. With a partial load factor of 1.5 × 1.1 and the spacing of beams at 3.46 m the ultimate force per beam is 1525 kN. This force is assumed to act parallel to the girders.

Other loading actions to be considered in detailed design are those from concrete shrinkage and differential temperature. Both these loads stress the concrete slab in a different way to the underlying steel beam, causing a force at the steel–concrete interface. The loadings are self-equilibrating and will not cause the structure to fail and so can generally be ignored at the ultimate limit state. However, as connector design is carried out at the serviceability limit, they should at least be considered. Shrinkage and differential temperature will cause additional deflections, which again may need to be considered. Shrinkage of the slab on a simply supported beam results in longitudinal shears of an opposite sign to the shears from loading on the span. For the differential temperature case with the slab cooler than the beam, longitudinal shears are again of opposite sign to the general dead and live loads. Where the slab is hotter than the beam, positive shears can result; however, they are unlikely to be significantly larger than the permanent shrinkage shears and so for simply supported structures both shrinkage and temperature effects can usually be safely ignored for the interface design.

Analysis

The modelling of any bridge, even the simplest of single-span beam-and-slab layout, involves some generalisation. Provided the limitations of the models are recognised then a series of simple models is usually preferable to a more complex model that tries to represent all aspects of the structure. For a beam-and-slab deck the most popular form of analysis is the grillage; this models the bridge deck in only two dimensions, and represents only the shear, bending and torsional interactions in the structure. The limitations of this method are well known [45], but in general it provides sufficient accuracy for the design of the main girders for most bridges. The grillage method does not consider longitudinal-load effects. For the soil loading a separate line model is used; this could be carried out by hand methods but the use of a computer model with nodes at similar points to those in the grillage allows the superposition of results more easily. The grillage can also be used to determine slab moments and shears if the transverse members are proportioned to suit an effective strip method [39]. Some designers advocate the use of an influence surface [51] in combination with the grillage to design the slab; this complicates analysis and leads to a very conservative design. If more detailed slab models are to be considered, the methods should recognise the significant in-plane or arching effects [42, 46] that occur in these structures (see Chapter 5). If the slab span is less than 3.2 m then the simplified methods of taking arching into account [47] may be appropriate.

The grillage model layout for this bridge is shown in Fig. 3.4. Three grillage models are used, with (i) a non-composite beam and zero slab stiffness, (ii) short-term concrete properties for the composite sections, and (iii) long-term composite properties. The line beam was modelled conservatively assuming soil loads on the long-term composite section. The temperature fluctuation causing the soil pressures is composed of annual and daily fluctuations. The model could have considered part of the load applied to a short-term composite section; this additional complication is not worthwhile for such a small structure (it is more applicable to larger structures).

Table 3.1 shows the relevant section properties for the grillage (for calculation of section properties see Appendix B). It should be noted that the torsional properties of the beams are effectively those of the slab as the torsional stiffness of the girders is negligible. The torsional properties are halved to more accurately represent this form of structure [45] (for a structure of this form setting the torsional stiffness to zero and only considering shear and bending effects would also lead to a safe solution). The end diaphragms and intermediate bracing have been included in the grillage; this generally gives a less conservative estimate of forces than an

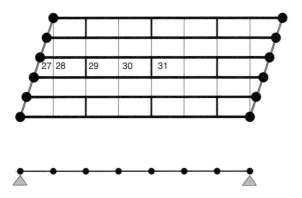

Figure 3.4 Models for integral bridge deck slab.

Table 3.1 Blythe Bridge section properties

Section	A: m^2	I: m^4	C_J: m^4	Z: m^3
Girder non-composite	0.070	0.036	0	0.052
Girder composite $n = 12$	0.100	0.088	0.0005	0.67
Girder composite $n = 6$	0.200	0.106	0.001	0.07
Bracing	0.006	0.006	0	–
Bracing $n = 12$	0.070	0.0115	0.0005	–
Bracing $n = 6$	0.120	0.0115	0.001	–
End wall	0.100	0.050	0.0004	–
Slab $n = 12$	0.064	0.0003	0.0005	–

alternative method, which is to not model these restraints, and later design them to accommodate the displacements of the grillage. As weathering steel is used the section properties used are those where a 1 mm corrosion allowance has been made on all exposed surfaces [49].

Weathering steel

Weathering steel has an enhanced resistance to corrosion compared to normal steel; the initial rust layer forms a protective barrier that slows dramatically further corrosion [48]. Typically an allowance of 1 mm to each exposed face should be allowed as a loss of section for a typical British structure with a 120-year design life [49]. For the interiors of box girders or other enclosed sections not subject to damp conditions this allowance may be reduced. In some locations the allowance may need to be increased. If the steel parts are continually wetted, are in coastal regions, or are in areas likely to be affected by de-icing salts the corrosion rates can be significant and weathering steels should not be used.

The cost of painting may be 11 to 18% of the steelwork cost (see Fig. 1.7) and the primary advantage of using weathering steel is to eliminate this. Furthermore, the need for future repainting is removed and future maintenance costs reduced.

Design

For this example three elements will be looked at in more detail: the girder bottom flange, the shear connectors and the deck slab at the interface with the girder. The moments, shears and axial loads from the analysis at the ultimate and serviceability limit state for a midspan section and a support section (nodes 31 and 27 on Fig. 3.4) are given in Table 3.2. The steelwork layout is shown in Fig. 3.5.

Assuming the section is to be checked as a non-compact section, the stress in the bottom flange at node 31 is:

$$f_{ab} = \frac{N_1}{A_1} + \frac{M_1}{Z_{b_1}} + \frac{N_2}{A_2} + \frac{M_2}{Z_{b_2}} + \frac{N_3}{A_3} + \frac{M_3}{Z_{b_3}} \tag{3.5}$$

$$f_{ab} = 0 + 121 + 0 + 32 + 8 + 165 = 326 \,\text{N/mm}^2$$

where subscripts 1, 2, 3 etc. are the various stages of construction. For a steel of thickness 28 mm the yield stress f_y is 345 N/mm^2 giving a limiting stress $(0.95f_y)$ of 327 N/mm^2, and so the section is just satisfactory. Another way of expressing

Table 3.2 Blythe Bridge analysis results

Node No.		SLS			ULS		
		M: MNm	V: MN	N: MN	M: MNm	V: MN	N: MN
Non-composite	27	0	0.55	0	0	0.60	0
	29	3.6	0.27	0	4.0	0.30	0
	31	5.2	0	0	5.7	0	0
Composite $n = 12$	27	0	0.15	0.93	0	0.22	1.4
	29	1.1	0.07	0.93	1.6	0.10	1.4
	31	1.35	0	0.93	1.9	0	1.4
Composite $n = 6$	27	0	1	0	0	1.3	0
	29	5.4	0.75	0	7.5	1	0
	31	7.5	0.35	0	10.5	0.45	0

Equation (3.5) is:

$$\frac{M}{M_D} + \frac{N}{N_D} < 1 \qquad (3.6a)$$

where

$$M = M_1 \frac{Z_{b_3}}{Z_{b_1}} + M_2 \frac{Z_{b_3}}{Z_{b_2}} + M_3 \qquad (3.6b)$$

$$N = N_1 \frac{A_3}{A_1} + N_2 \frac{A_3}{A_2} + N_3 \qquad (3.6c)$$

These equations are preferred by the author, as rather than checking stresses at each node the moments and axial forces can be modified by an additional section factor during the analysis with the partial load factors; the moment diagrams can then be plotted directly and the M_D diagram superimposed. For the simply supported beam example there is little difference in the number of calculations to be carried out; however, for a more complex structure the use of spreadsheets to calculate stresses can be virtually eliminated. The use of a graphic method of design also aids reviews and checking, which is more difficult with tabulated stresses. The form of Equation (3.6a) is also used where axial loads are more predominant (see Chapters 8 to 11).

The longitudinal-shear flow is calculated at the serviceability limit state from the vertical shear (Equation (1.16)); it is a maximum at the support (node 27) and again is summed at all stages.

$$Q_l = Q_{l_1} + Q_{l_2} + Q_{l_3} \qquad (3.7)$$

From the section properties the Q_l/V ratio is known at each stage (Appendix C).

$$Q_l = 0 + 0.46 \times 0.17 + 0.53 \times 1.1 = 0.66 \, \text{MN or } 660 \, \text{kN}$$

Using 19 mm diameter studs with $P_u = 109 \, \text{kN}$ (Table 1.5) and Equation (1.17b):

$$No = \frac{Q_l}{0.55 P_u} = \frac{750}{60} = 12.5$$

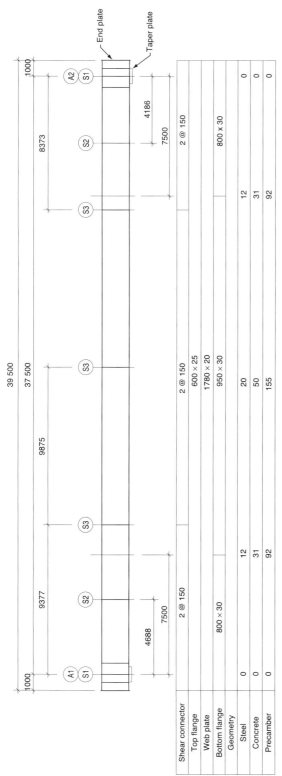

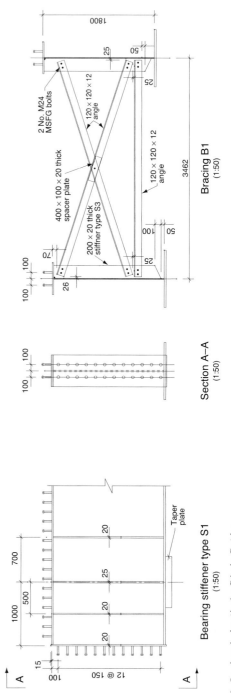

Figure 3.5 Steelwork details for Blythe Bridge.

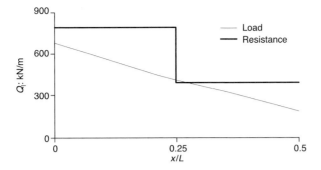

Figure 3.6 Longitudinal-shear diagram, with connector requirements.

The connectors required are calculated at the other nodes and plotted as shown in Fig. 3.6. Two connectors at 150 mm centres give sufficient capacity at the support; the connectors could be curtailed between nodes 28 and 29 to two at 300 mm centres. In general, connector spacing should be kept constant for at least 10% of the span. At the supports the connection capacity should exceed the applied shear, at other locations the shear flow can locally exceed capacity by up to 10%.

Having confirmed the shear plane is satisfactory through the connectors the longitudinal-shear capacity of the concrete slab should be checked using Equations (1.24) and (1.25). At node 27 (at the support) the shear flow at the ultimate limit state is 890 kN/m. For a pair of 150 mm-high connectors 200 mm apart the length of plane y–y is 520 mm, the length of plane z–z (see Fig. 1.9) is 330 mm (ignoring the lower 60 mm of the slab where the plane runs along the edge of the precast formwork), using Equation (1.25a).

The maximum longitudinal shear $Q_{1\,max} = 0.15 L f_{cu} = 0.15 \times 330 \times 40 = 1980$ kN/m, for z–z the shortest plane. This is greater than the applied shear and is satisfactory. Plane y–y has the least reinforcement with two T12 bars at 300 mm centres running between the permanent formwork, giving an area of reinforcement of 747 mm^2/m each time the reinforcement crosses the shear plane. Using Equation (1.24):

$$Q_1 = 0.9 L_s + 0.7 A_s f_y = (0.9 \times 520) + (2 \times 0.7 \times 747 \times 460 \times 10^{-3})$$

$$= 948 \, \text{kN/m}$$

This is again greater than the applied shear. Both shear planes are satisfactory.

The detailing rules [4] for reinforcement through the shear plane require that 50% of the bars are within 50 mm of the connector base; to achieve this one of the two bars passing between the precast units should be fixed low as shown in Fig. 3.7.

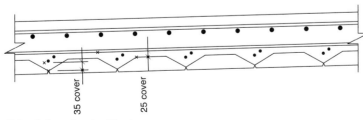

Note: all dimensions in millimetres

Figure 3.7 Detailing of reinforcement at precast permanent formwork.

Compact sections

So far we have considered beams using elastic analysis and elastic section properties. Where sections meet the outstand limits and connector spacing requirements for compact sections (see Table 1.3), the increased plastic section properties may be used (Appendix D). For steel beams the plastic modulus is typically 10 to 20% higher than the elastic modulus. For steel concrete composite sections with the concrete slab element in compression the plastic modulus is approximately 5 to 20% higher, but will depend on the modular ratio of the section. A primary advantage of using a compact or plastic section is that all loads may be assumed to act on the final section, such that Equation (3.6a) may be used, with Equations (3.8a) and (3.8b):

$$\frac{M}{M_D} + \frac{N}{N_D} < 1 \tag{3.6a}$$

where

$$M = M_1 + M_2 + M_3 \tag{3.8a}$$

$$N = N_1 + N_2 + N_3 \tag{3.8b}$$

However, to take advantage of this additional bending capacity the shear connection should be designed to carry the entire shear:

$$Q_l = (V_1 + V_2 + V_3)\frac{Ay}{I} \tag{3.9}$$

Compact section design therefore tends to allow smaller sections for bending but higher shear connector requirements.

Portal frame structures

Integral portal structures are formed from the composite deck structure being connected to the full-height abutment. The advantage of this form is that the moments are nearer those of a fixed-end beam rather than simply supported (Fig. 3.8); deflections are also significantly reduced. A disadvantage, structurally, is that a larger moment range has to be considered due to the variation of soil pressures, which range from low (active) to high (almost passive).

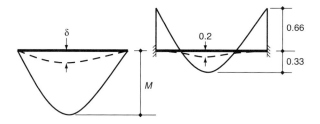

Figure 3.8 Comparison of moments and deflections for fixed and simply supported structures.

Example 3.2

The second example in this chapter is a fully integral portal structure; the bridge was an alternative design proposal for a new road in Staffordshire. The structure is similar to that of Fig. 3.1(e) and has a 22.5 m-span multiple-rolled universal-beam bridge deck supported on high-modulus steel piles, the abutment being part of a longer retaining wall system.

Loads and analysis

The soil loads on the wall will vary from low active pressures (K_a) to the larger integral pressures (K^*). The maximum and minimum soil pressures are shown in Fig. 3.9(a). For a wall of this form the high integral pressures occur in the top half to two thirds of the wall height. They may be assumed to be back to at-rest (K_o) near the front ground line, this distribution being dependent on the relative stiffness of the wall and soil. The characteristic soil loads are approximately 4.5 MN maximum (K^*) and 0.7 MN minimum (K_a).

Characteristic deck loading of 0.5 MN, 0.225 MN and 1.3 MN for dead, surfacing and live loads, respectively, are also assumed. Analysis of the structure is carried out as a portal frame strip, and the results are summarised in Fig. 3.9(b). The maximum sagging moment at midspan is in combination with the minimum soil loads on the wall; the maximum hogging moment is in combination with the maximum soil loads. The moment at the portal knee is similar to the maximum midspan moment but causing tensions in the concrete elements (see Chapter 4).

To reduce the effects of the range of maximum and minimum soil pressures the walls may be inclined. If the wall of the portal is leaning at an angle near to the soil's shear resistance, the range of pressures will be minimised, leading to an inclined portal or arch form as the best portal shape.

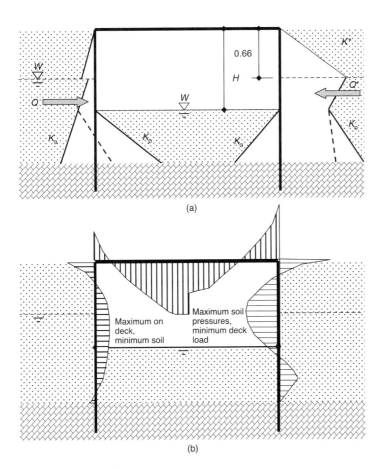

Figure 3.9 Example 3.2: (a) maximum and minimum soil pressures; (b) bending moments on the portal.

Effects of skew

Examples 3.1 and 3.2 have only a small skew; for bridges with a larger skew angle (θ) the soil forces (Q_s) are offset (Fig. 3.10(b)). If the eccentricity (e) of the forces becomes too large some rotational movement of the abutment may occur. For stability a shear force will occur along the abutment:

$$V = Q_s \frac{e}{L} \qquad (3.10)$$

where e increases with the skew:

$$e = L \tan \theta \qquad (3.11)$$

The limiting value of V is estimated from the friction on the abutment, assuming a factor of safety from sliding of 1.5:

$$V_{max} = \frac{Q \tan}{1.5}(0.67\phi) \qquad (3.12)$$

where ϕ is the soil angle of shearing resistance; a value of 0.67 is used as the failure surface is likely to be at the interface with the wall. With $\phi = 42°$, $V = V_{max}$ at approximately 28°. Hence for skews above this value the behaviour of the structure should be considered carefully.

For a high-skew crossing the arrangement of the beams could be modified such that they form a square span with no eccentricity; this gives a larger deck area than the skewed bridge (Fig. 3.10(c)). For bridges where the width is greater than the span and the girders are square with only skewed edge beams (Fig. 3.10(d)), the maximum soil pressures are higher adjacent to a–b and c–d. The limiting skew angle increases as the width-to-span ratio (B/L) of the deck increases (Fig. 3.10(e)).

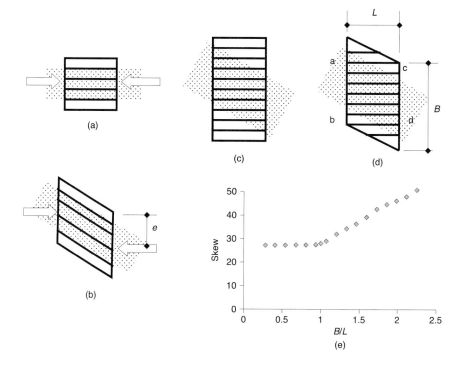

Figure 3.10 Beam layouts for integral bridges: (a) square bridge with no load eccentricity; (b) narrow skew bridge with eccentricity of load; (c) over-wide bridge; (d) wide bridge with skew ends; (e) limiting skew angles for integral bridges with various width-to-span ratios (B/L).

Example 3.3

It is possible to design integral bridges outside the range outlined in Fig. 3.10. The final example in this chapter is the Nanny Bridge [50], where a dual carriageway crosses a river at a skew of 48° (Fig. 3.11). To overcome the issues of skew, this design is formed using a portal beam arrangement with reinforced-earth walls. The portal is fully fixed at one end with movement occurring at only one end. Only the top of the portal is in contact with the soil, limiting the height in which the soil–structure interaction is accommodated. The fixing at one end does, however, limit the length of structure that can be used to about half of that of a conventional integral structure (40 m rather than 80 m). The portal also allows the beam to be profiled, an aesthetic requirement stipulated for this structure.

At the portal end the tensile load is initially carried by only the steel flange. There is some strain incompatibility with the concrete which has no load. Where a force is applied to only one element of a composite section the force cannot be transferred instantaneously, it must spread over a finite length (L_s). The length over which the applied force spreads is dependent on the stiffness of the connector between the two elements and can be estimated from:

$$L_s = 2\left(K\frac{\delta N}{\delta \varepsilon}\right)^{0.5} \tag{3.13}$$

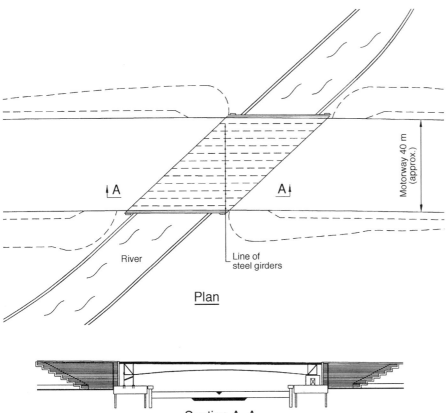

Plan

Section A–A

Figure 3.11 Details of the Nanny Bridge.

where $K = 0.003 \, \text{mm}^2/\text{N}$ for stud connectors or 0.0015 for other connectors, and $\delta\varepsilon$ is the difference in free strain between steel and concrete.

At the end of the beam the connector requirement is calculated from the longitudinal shear and the additional force that is required to transfer the loads from the steel to the concrete. The shear at the end of the beam for this example is 1.4 MN, and Q_1/V is 0.34 (Appendix C):

$$Q_1 = 0.34 \times 1.4 = 0.48 \, \text{MN/m}$$

The force in the tensile stiffener and flange at the joint with the all-steel portal is 4.4 MN; only about 35% of this force will transfer into the slab, $\delta N = 4.4 \times 0.35 = 1.54 \, \text{MN}$.

The flange has an area of $0.225 \, \text{m}^2$, its stress is: $44/0.225 = 195 \, \text{N/mm}^2$

$$\varepsilon_a = \frac{f}{E_a} = \frac{195}{210\,000} = 0.00093, \qquad \varepsilon_c = 0$$

Thus $\delta\varepsilon = \varepsilon_a - \varepsilon_c = 0.00093$. Using Equation (3.13):

$$L_s = 2\left(\frac{K\delta N}{\delta\varepsilon}\right)^{0.5} = 2\left(0.003 \times \frac{1.54}{0.00093}\right)^{0.5} = 4.46 \, \text{m}$$

$$Q_{l'} = \frac{\delta N}{L_c} = \frac{1.54}{4.46} = 0.35 \, \text{MN}$$

The total longitudinal shear is $0.48 + 0.35 = 0.83 \, \text{MN/m}$ or $830 \, \text{kN/m}$.

An alternative way of estimating L_s is to assume a $1:2$ spread from the end of the beam to the midspan of the slab. The beam is non-composite at the start of spread and fully composite beyond L_s, with the force being transferred uniformly over this length. For the Nanny Bridge using this method $L_s = 3.4 \, \text{m}$, smaller than the previous estimate.

Aesthetic requirements on this structure dictated that, rather than using a weathering steel, the steel would be painted. For the Nanny Bridge the steelwork for the girders was provided with a $450 \, \mu\text{m}$ minimum thickness glass flake epoxy in the fabrication shop with a polyurethane finishing coat applied on site. For the portal legs, which could be submerged in flood conditions, an extra 1 mm of steel was allowed all around together with an aluminium spray and a $350 \, \mu\text{m}$ multi-coat paint system.

Painting

The corrosion of steel is an electrochemical process where the steel in the presence of oxygen and water converts to a hydrated ferric oxide or rust. The rate of corrosion will be heavily influenced by the amount of time the steelwork is wet; if the bridge can be detailed such that the steel section is protected then the rate of corrosion will be reduced. Atmospheric pollutants such as sulphates and chlorides will also have an effect on the corrosion rate, consequently industrial and coastal regions will have higher corrosion rates. The steel is normally painted to protect it from corrosion. The effectiveness of the paint system will depend upon the surface condition of the steel, the paint or protective system used, the procedures used to apply the protective system and the environment in which it is applied.

The performance of a protective system is highly influenced by its ability to adhere properly to the steel. Mill-scale, rust, oil, water or other surface contamination of the steel will need to be removed. Blast cleaning is usually used to clean the steel. The cleaning also creates a rough surface profile that helps the adhesion of the paint system. The process used to apply the paint and the environmental conditions when applied are also significant to the adherence of the paint. Most specifications for new bridges require the surface preparation and the first coats of metal and paint to be applied at the place of fabrication, where the environmental conditions can be controlled and the paint spray applied. Site brush applied paint systems are usually to be avoided except for the final finishing coat.

The protective system used will depend upon the environment, access conditions and the intended life of the paint until first maintenance. Bridges in coastal regions or with difficult access will generally require a better (and usually more expensive) system than inland structures with better access. Table 3.3 outlines some common systems for highway and railway structures. The systems use paint or a metal coating system with paint. The metal coating is a thermal sprayed aluminium coating.

Paint systems consist of three basic stages: a primer applied directly to the steel or to the sealed metal coating, the main intermediate stage which may be one thick coating or several thinner layers and finally the finishing coat. The choice of intermediate coating will depend on the environment and to some degree the fabricator applying the system. Glass flake epoxy coating has the advantage of building up a thick layer in a single application, whereas micaceous iron oxide epoxy is applied in three or four layers. The outer layer of paint is the layer that provided the finish to the structure, particularly its final colour.

The final colour of the bridge significantly affects its character. Battleship grey has been a common colour particularly for motorway bridges, meaning they look very like concrete structures. The use of a dark grey or black paint can enhance the shadow effects and draw the eye to the thinner parapet coping (Fig. 4.1). Earthy green and brown colours are used particularly in rural areas; a deep blue can also blend in well. Brighter colours such as yellow or red can be used if the structure is spectacular. White or silver is popular, modern and fashionable; however, the bridges can become dirty very quickly and therefore require a good

Table 3.3 Protective systems for bridges

Environment/ access	Preparation	1st coat	2nd coat	3rd coat	4th coat	Thickness: μm	Relative cost
Protected (interior of box)	Blast clean	Zinc epoxy primer	MIO			200	0.85
Inland with good access	Blast clean	Zinc epoxy primer	MIO	MIO	Polyurethane finish	300	1
Inland with poor access	Blast clean	Epoxy primer	Glass flake epoxy	Polyurethane finish		450	1.25
Marine or industrial	Blast clean, aluminium spray	Epoxy sealer	Zinc epoxy primer	MIO	Polyurethane finish	400	1.6

MIO, micaceous iron oxide.

cleaning and maintenance regime. The designer should take great care in the choice of colour; it is a major factor in the layperson's view of the structure.

Shrinkage

In Example 3.1 it was noted that for a simply supported span the effects of concrete shrinkage could be ignored as the effects were of different signs. For a cantilever or the hogging area of a beam the effect of shrinkage is additional to the bending effects (Fig. 3.12(b)).

For a deck slab the shrinkage strain is estimated from Fig. 1.2(b) and is typically 0.0001 to 0.0002. If restrained this shrinkage would induce a tensile force in the slab of:

$$G_{cs} = \varepsilon_s E_{c'} A_c \qquad (3.14)$$

In the actual structure there will not be a rigid restraint and the force will be shared between the beam and slab. The proportion of the force carried by the beam or slab is dependent on its stiffness compared to that of the whole composite section:

$$G_a = G_{cs} \frac{A_a}{A_{ac}} \qquad (3.15)$$

The force has to be transferred into the steel by shear connectors; this will occur near the ends of the beam. The length over which the force is transferred will depend on the connector stiffness and is estimated using Equation (3.13). The longitudinal shear flow (Q_l) is assumed to vary from a maximum at the end of the girder to zero at L_s from the end:

$$Q_l = 2 \frac{G_a}{L_s} \qquad (3.16)$$

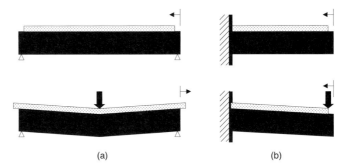

Figure 3.12 Relative free movement of slab: (a) simply supported with shrinkage and under loading; (b) cantilever with shrinkage and under loading.

(a)   (b)

Differential temperature

Two temperature distributions across a girder are considered, a positive temperature difference with the slab hotter than the girder, and a reverse temperature difference. These temperature distributions induce forces into the slab and girder, the reverse temperature difference having an effect similar to that of shrinkage.

For Example 3.3, which is restrained rotationally at the ends producing a significant fixing moment (Fig. 3.8), the effects of shrinkage and differential shrinkage will be considered. The shrinkage strain for a slab in a damp environment is estimated as 0.00015 (Fig. 1.2). The concrete modulus from Equation (1.3) is:

$$E_{c'} = 0.5E_c = 0.5 \times 34\,000 = 17\,000 \, \text{MN/m}^2$$

giving a modular ratio n of 12.

The section properties are given in Appendices B and C. The steel area $A_a = 0.093 \, \text{m}^2$, $A_c = 0.765 \, \text{m}^2$, $A_{ac} = 0.157 \, \text{m}^2$.

Using Equation (3.14) the restrained slab force is:

$$G_{cs} = \varepsilon_s E_{c'} A_c = 0.00015 \times 17\,000 \times 0.765 = 1.95 \, \text{MN}$$

Using Equation (3.15) the force transferred to the steel girder is:

$$G_a = G_{cs} \frac{A_a}{A_{ac}} = 1.95 \times \frac{0.093}{0.157} = 1.15 \, \text{MN}$$

Similarly for differential temperature the force to be transferred is 0.38 MN.

For this structure the transfer length has already been calculated to be $L_s = 4.46 \, \text{m}$. Using Equation (3.16) the maximum longitudinal shear is:

$$Q_l = \frac{2G_a}{L_s} = 2 \times \frac{(1.15 + 0.38)}{4.46} = 0.69 \, \text{MN/m}$$

This force is added to that previously calculated. Where critical, the effects of shrinkage and differential temperature will increase the connector requirement locally at the ends of the bridge.

4

Continuous bridges

...the key benefit of continuity is a reduction (or elimination) of bearings and joints...

Introduction

The bridges considered so far have been single-span structures. For many bridges two or more spans are required. If a bridge is required to span a 50 m-wide motorway it is often more economic to use two 25 m spans rather than a single 50 m span. For most bridges of two or more spans it is now fairly standard to use a continuous bridge, with full continuity over intermediate piers, unless there are significant settlement issues that require the flexibility given by a series of simply supported spans. For these structures the continuity over the intermediate piers produces a hogging moment, inducing a tension in the concrete slab and compression into the lower flange of the bridge. This reduces the efficiency of the structure, because concrete is poor in carrying tension and must have additional reinforcing steel added; the steel flange in compression is unrestrained by the concrete slab and more prone to buckling. There are also further complications in that high shears occur at the intermediate supports together with large moments, which require a consideration of shear–bending interaction.

Continuous bridges tend to be stiffer than simply supported bridges of the same span, and so the span-to-depth ratios can be increased (Table 4.1) compared to simply supported structures (Table 2.1). Overall there is little difference in the

Table 4.1 Typical span-to-depth ratios for continuous bridges

Element	Span-to-depth ratio	Remarks
Constant depth main steel girders (highway bridges)	20 to 25 16 to 20	Internal spans End spans
Haunched main girders (highway bridges)	20 to 30 12 to 20	Midspan At supports
Transverse beams for ladder-beam type (highway bridges)	10 to 22	
Constant depth main steel girders (rail bridges)	10 to 20 8 to 18	Internal spans End spans
Haunched main girders (rail bridges)	15 to 20 9 to 15	Midspan At supports
Transverse beams for ladder-beam type (rail bridges)	5 to 20	

Figure 4.1 Continuous steel–concrete bridge spanning the M5 motorway [14].

cost of steelwork between a continuous structure and a series of simply supported spans; the key benefit of continuity is a reduction (or elimination) of bearings and joints.

Figure 4.1 shows a typical two-span bridge over the M5 motorway to the west of Birmingham [14] with continuity across the spans. This motorway was widened with the new bridges being primarily steel–concrete composite structures, which were relatively quick and simple to erect over the existing road, causing minimum disruption to traffic using the motorway.

Motorway widening

When constructing a new motorway it has been a common practice to provide structures that simply span the proposed highway. Significant growth in traffic on motorways in recent years has meant that many have had to be widened to increase capacity. Given the relatively short period since the original construction it can be argued that providing over-long structures to allow for future road widening would be beneficial. The benefit of this can be demonstrated by the use of a net present value calculation, where the values of the future costs (C_f) are discounted at a chosen rate (i) over a number of years (N) to the net present value (C_{np}):

$$C_{np} = \frac{C_f}{(1+i)^N} \tag{4.1}$$

Consider for example a motorway bridge with an initial cost of £800 000 for the minimum span or £1 000 000 for an over-wide layout. With a discount rate of 6% [71] and a time to widening of 25 years the additional cost for building the minimum span and rebuilding at a future date using Equation (4.1) is:

$$C_{np} = \frac{C_f}{(1+i)^N} = \frac{1\,000\,000}{(1+0.06)^{25}} = £233\,000$$

So the effective cost of the minimum span plus future rebuilding is £800 000 + £233 000 = £1 033 000, which is similar to the cost of building with provision for widening. This offers only a rough guide; the chosen discount rate and time to rebuilding will significantly affect costs. The issues of costs of delays are also ignored; for major roads these additional delay costs can be significant and may be more than the rebuilding cost of the structure [60].

Motorways may be widened in a number of ways, using symmetric or asymmetric widening, by parallel widening or by the addition of feeder roads. In symmetric widening the hard shoulders of the motorway are reconstructed to form a new lane and additional shoulders constructed. Asymmetric widening reconstructs the hard shoulder and extends the carriageways on one side only. Both these methods tend to cause significant disruption to the existing carriageways. The parallel widening technique involves the construction of a complete new carriageway adjacent to the existing one and, when this is complete, modifications to the existing carriageway can be carried out. This method requires more land and is not suitable in urban locations but causes less disruption to road users. The fourth method, constructing additional feeder roads adjacent to the existing road, causes the least disruption to traffic but can only be used if some segregation of traffic is accepted.

Minimising the disruption to traffic on the motorway is a key element of motorway widening, and the rebuilding of bridges requires methods that facilitate this [53]. Steel–concrete composite bridges are one of the best ways of achieving this as the steelwork can be prefabricated off site and lifted in during night-time closures of the motorway. The layouts of the reconstructed bridges will be different for each widening type, their character affected by the form of widening (Fig. 4.2). For the symmetric and asymmetric widening a single-span structure is favoured, in order to minimise disruption during the construction of a pier. For widening to four or five lanes the spans become long and it may be more economic to

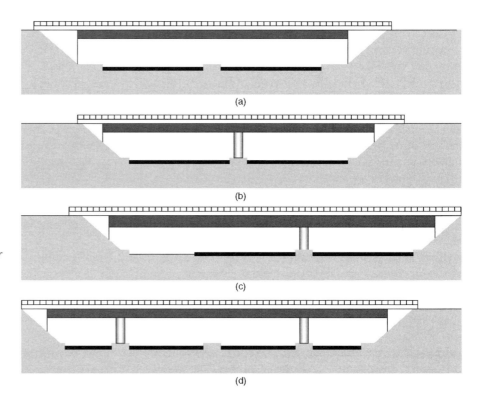

Figure 4.2 Bridge spans for motorway widening: (a) single span; (b) two span, both for use with symmetric or asymmetric widening; (c) two asymmetric spans for parallel widening; (d) multi-span for feeder roads.

48

provide a central pier and accept some small additional traffic disruption. For parallel widening an asymmetric two-span layout is required to allow the new bridge to span the existing carriageways prior to its reconstruction or removal. The feeder road option gives a multi-span form.

Moment–shear interaction

The maximum design bending capacity (M_D) of a beam is derived from its section properties including the web. In Examples 2.1 and 3.1 the web accounts for about 10 to 20% of the bending capacity (see also Appendix A). The maximum design shear capacity (V_D) of a beam is derived from the section including the flanges. Figure 1.6 shows that there is an increase in shear capacity for a beam with stiff flanges, particularly at small panel ratios. The moment–shear interaction occurs primarily at the onset of tension field action in the web [54], where some local web buckling occurs and the shear is carried as a diagonal tie. If the flange is stiff the width of the tension field is larger and the shear capacity increased; the failure in shear involves a local bending failure in the flange.

If the beam bending capacity is derived only using the flanges such that the web is ignored (M_f), then there will be no interaction and the full shear capacity of the web can be utilised. If the shear capacity is derived from the web (V_w) assuming no flange interaction (with $m_f = 0$ in Fig. 1.6) then the full moment capacity of the section can be carried. This simple interaction is shown in Fig. 4.3(a). For multi-beam bridges of 20–40 m span the difference between M_f and M_D or V_w and V_D is small and the simple interaction will not lead to significant conservatism. Where the girder webs and flanges are more substantial the simple interaction may lead to conservatism and a full interaction curve (as Fig. 4.3(b)) may need to be considered. For the full interaction (Equations (4.2a and b)) it is assumed that full bending can be carried at half V_w and that full shear can be carried at half M_f.

$$\frac{M}{M_D} + \left(1 - \frac{M_f}{M_D}\right)\left(\frac{2V}{V_w} - 1\right) < 1 \tag{4.2a}$$

$$\frac{V}{V_D} + \left(1 - \frac{V_w}{V_D}\right)\left(\frac{2M}{M_f} - 1\right) < 1 \tag{4.2b}$$

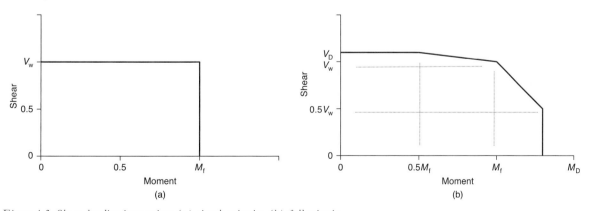

Figure 4.3 Shear–bending interaction: (a) simple criteria; (b) full criteria.

Example 4.1

The first example of this chapter is another bridge on a motorway around Birmingham near Example 3.1; the new motorway was constructed to relieve traffic on the original motorway where there was little room for widening. This bridge spans the motorway on two spans of 37.55 m and 31.34 m and is continuous across the pier. The bridge is semi-integral and supported on a reinforced earth wall with a bank seat (Fig. 4.4). The bridge is 11.8 m wide and consists of four constant depth girders curved in plan to follow the highway. As before, the bridge has been sized for 30 units of HB and HA loading.

Loads and analysis

The loads imposed upon the bridge are determined as in the previous examples. The modelling of the bridge is carried out as a grillage and a line beam, to estimate deck and soil loads in a similar manner to Example 3.1. However, as the bridge is continuous some complications arise at internal supports. Two methods of analysis are available:

Method 1 Analyse the bridge using composite section properties assuming the full concrete area, even over the pier. The results of the analysis are used to estimate

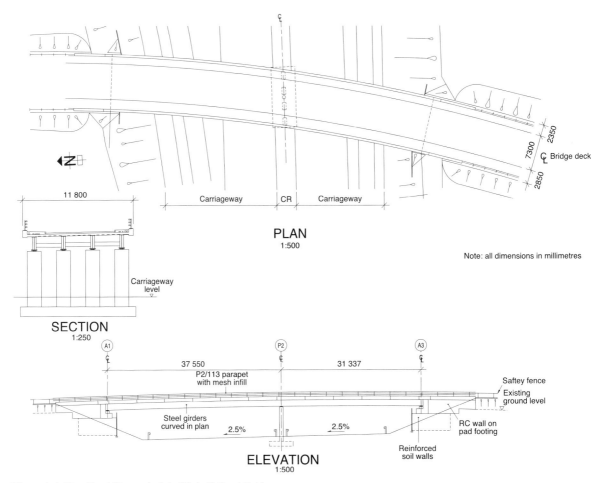

Figure 4.4 Details of Example 4.1, Walsall Road Bridge.

the tensile stresses in the concrete. If the concrete tensile stress exceeds $0.1f_{cu}$ then the section is assumed to have cracked and the sagging moments from the analysis are increased by 10%; no reduction in hogging moment is permitted.

Method 2   Calculate section properties of midspan sections as above but at the pier and approximately 15% of the span to each side use section properties that ignore any concrete in tension (any reinforcement in the slab should be considered, see Appendix B).

Method 2 is most commonly used because with current computing methods it is relatively simple to use; generally it will give slightly lower design moments than Method 1.

For Example 4.1, Method 2 is used. The moment and shear diagrams for analysis are shown in Fig. 4.5. The moment envelope has been factored taking into account the relative section moduli as outlined in Chapter 3, Equation (3.6b). It should also be noted that the peak moments at the supports have been rounded.

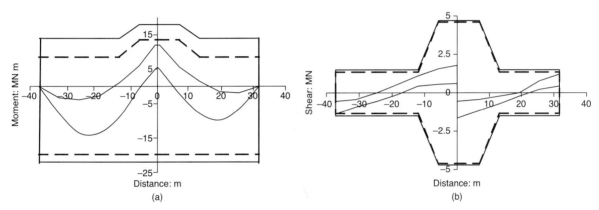

Figure 4.5   Analysis results: (a) bending moments; (b) shear.

Moment rounding

Moment rounding is common for concrete structures [55] where a spread of the load from the bearing plate to the neutral axis can be visualised (Fig. 4.6(a)). For a steel girder where the compression path has to follow stiffeners to prevent the buckling of plates, the spread of load is less clear. If a single-leg stiffener is used then rounding will be small, for a multi-leg stiffener there is a greater spread of load (Fig. 4.6(b)). The reduction in moment M' is dependent on the reaction and the width of spread.

$$M' = \frac{Ra}{8} \tag{4.3}$$

where R is the reaction, 3.3 MN in this example, and a is the length of the effective support, 1200 mm in this case.

$$M' = \frac{Ra}{8} = \frac{3.2 \times 1.2}{8} = 0.5\,\text{MN}\,\text{m}$$

The section properties of the girder have been calculated using the steel section (Fig. 4.7) and the two layers of T16 reinforcing bars at 150 mm centres in the

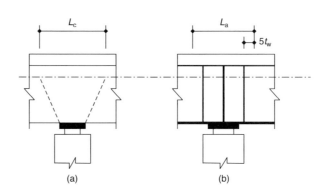

*Figure 4.6 Moment
rounding for: (a) a concrete
structure; (b) a steel
structure.*

slab (see Appendix B). The bottom flange modulus $Z_b = 0.058\,\mathrm{m}^3$, $Z_t = 0.045\,\mathrm{m}^3$ and the transverse radius of gyration of the steel section, $r = 372\,\mathrm{mm}$.

The bottom flange is in compression so the limiting compressive stress is to be estimated. For the completed bridge the slab is in place; this has a large lateral inertia preventing a lateral torsional buckling failure mode, the mode of failure being a more local mode involving bending of webs and stiffeners. Bracing at the pier and at 9.0 m either side of the pier is provided. Using the basic lateral torsional buckling slenderness parameter (λ) as a conservative estimate of behaviour

$$\lambda = k_1 k_2 k_3 \frac{L_e}{r} \tag{4.4a}$$

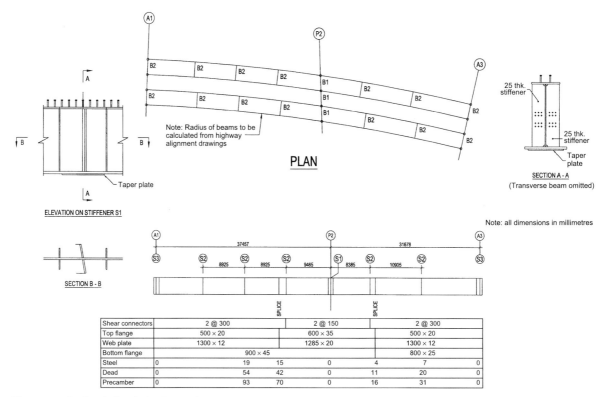

Figure 4.7 Steelwork details for Example 4.1.

The coefficient k_1 is taken as 1.0 because the girder is a fabricated section. The moment drops significantly from the support to the bracing, meaning k_2 is less than 1; the compression flange has a transverse inertia less than the deck slab so k_3 is greater than 1. For moderate spans, at the support the equation simplifies to:

$$\lambda = \frac{L_e}{r} \tag{4.4b}$$

For this example the distance from the support to the effective brace is 9 m, using Equation (4.4b):

$$\lambda = \frac{L_e}{r} = \frac{9000}{372} = 24$$

From Fig. 2.6, at this value (less than 45) buckling is not an issue and the full moment can be carried. Using Equation (1.8): $M_D = 0.95 f_y Z_b = 0.95 \times 335 \times 0.058 = 18.5$ MN m for the bottom flange. The design resistance moment $M_D = 0.87 f_{ys} Z_t = 0.87 \times 460 \times 0.045 = 18.0$ MN m for the reinforcement (note the lower material factor normally used for the reinforcing steel). The bending capacity of the beam ignoring the web is:

$$M_f = 0.95 f_y A_f D \tag{4.5}$$

where D is the distance between flanges. The design resistance moment of flanges $M_f = 0.95 f_y A_f D = 0.95 \times 335 \times 0.0405 \times 1.450 = 20.7$ MN m for the bottom flange, and $M_f = 0.95 f_y A_f D_f + 0.87 f_{ys} A_s D_s = 318 \times 0.021 \times 1.325 + 400 \times 0.00837 \times 1.340 = 13.3$ MN m for the top flange and reinforcement. The tensile capacity of the composite top flange is the lowest and most critical. The moments M_D and M_f at the support and midspan are plotted on Fig. 4.5(a). It can be seen that at all locations the applied moment is less than the resistance moment.

For the web $d/t = 1285/20 = 64$. The web panel aspect ratio $a/d = 8500/1285 = 6.6$. The relative flange stiffness $m_f = bt_f^2/2d^2 t_w = 450 \times 45^2/2 \times 1285^2 \times 20 = 0.014$. Using Fig. 1.6, and interpolating between panel aspect ratios of 2 and 10, $V_D = 0.94 V_y$ and $V_w = 0.92 V_y$. Using Equations (1.9) and (1.10):

$$V_y = 0.55 f_y t_w d = 0.55 \times 355 \times 1.285 \times 0.020 = 5.0 \text{ MN}$$

$$V_D = 0.94 \times 5.0 = 4.7 \text{ MN}, \quad V_w = 0.92 \times 5.0 = 4.6 \text{ MN}$$

These values together with those for the midspan section are drawn on Fig. 4.5(b). It can be seen that at all locations the applied shear is less than the shear resistance. From Fig. 4.5 it is clear that $M < M_f$ and $V < V_w$ and so no $M-V$ interaction check is required.

Cracking of concrete

Over a pier support of a continuous beam the deck slab is in tension. If the stress in the concrete exceeds its limiting tensile strength then cracking will occur and the force in the concrete will be transferred to the reinforcement. For structures built in stages with the concrete initially carried by the steel section then the stresses in the reinforcement will be lower than the stresses in the steel section (Fig. 4.8).

The amount of and size of cracks in the slab will depend on the stress in the reinforcement. At the serviceability limit state the limiting crack width (w) to

UK codes [3] is:

$$w = 3a_{cr}\varepsilon_m \tag{4.6}$$

where a_{cr} is the distance from the reinforcement to the point where the crack width is considered (Fig. 4.8) and ε_m is the strain at that point. For design this may be rearranged to give the limiting stress:

$$f_{sc} = \frac{E_s w}{1.6 s_b} \tag{4.7a}$$

For European standards [7] the stress is limited in a similar way, for bars at a spacing (s_b) of 200 mm or less the limiting stress is:

$$f_{sc} = 200 - \frac{0.4 f_{ct} b h}{\alpha A_s} \tag{4.7b}$$

where f_{ct} is the tensile strength of concrete and α is the ratio of the area and second moment of area (AI) of the composite and non-composite section:

$$\alpha = \frac{(AI)_{a-c}}{(AI)_a} \tag{4.8}$$

The standards also recommend a minimum amount of reinforcement:

$$A_s = 0.9 k_c \frac{f_{ct}}{f_s} b h \tag{4.9}$$

The stress f_s is dependent on the bar diameter, smaller bars giving higher stresses. For bars of 20 mm diameter or less the ratio of f_{ct}/f_s can be taken as 0.01. For most beam and slab bridges k_c is 1.0.

For Example 4.2 using Equation (4.9) with $b = 3100$ mm and $h = 225$ mm:

$$A_s = 0.9 k_c \frac{f_{ct}}{f_s} b h = 0.9 \times 1 \times 0.01 \times 3100 \times 225 = 6277 \text{ mm}^2$$

Using T16 bars at 150 mm centres, the area provided is 8308 mm^2 and so is more than the minimum requirement. Using Equation (4.7a) with a crack width of 0.15 mm:

$$f_{sc} = \frac{E_s w}{1.6 s_b} = \frac{210\,000 \times 0.15}{1.6 \times 150} = 131 \text{ N/mm}^2$$

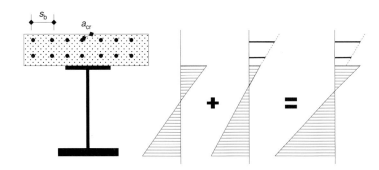

Figure 4.8 Build-up of stresses in a composite section in hogging regions of a deck; stresses in the reinforcement are usually lower than those of the steel girder.

Using Equation (4.7b) with $f_{ct} = 2.5\,\text{N/mm}^2$ and α from Equation (4.8), and using the properties from Appendix C:

$$\alpha = \frac{(AI)_{a-c}}{(AI)_a} = \frac{(0.087 + 0.008) \times 0.034}{0.087 \times 0.029} = 1.29$$

$$f_{sc} = 200 - \frac{0.4 f_{ct} b h}{\alpha A_s} = 200 - \frac{0.4 \times 2.5 \times 3100 \times 225}{1.29 \times 8308} = 135\,\text{N/mm}^2$$

As would be expected similar limiting stresses are obtained for both methods.

Bearing stiffeners

At the pier a high reaction is transferred from the girder to the fixed bearing via a stiffener. The stiffener is required to carry this and prevent buckling of the web. If the stiffener thickness (t_s) is greater than $d/60$ and the stiffener to each side of the web has an outstand ratio of $10t_w$ or less then overall and local buckling of the stiffener will not occur and the squash load of the section can be used:

$$N_s = (A_s + A_w)0.83 f_y \tag{4.10}$$

where A_s and A_w are the area of the stiffener and the web acting with the stiffener ($32t_w$, see Fig. 4.8). Assuming a 250 mm by 25 mm stiffener each side of the 20 mm web:

$$A_s = 12\,500\,\text{mm}^2$$

$$A_w = 12\,800\,\text{mm}^2$$

The axial load $N = (12\,500 + 12\,800) \times 0.83 \times 345 = 7\,300\,000\,\text{N}$ or 7.3 MN. This is significantly more than the applied reaction of 3.3 MN. The large stiffener is used to give sufficient room to bolt on the cross-beams stabilising the system. To allow future jacking for bearing replacement additional stiffeners are located each side of the main stiffener; the jacking stiffeners are 20 mm thick.

Precamber

Bridges deflect under load; there are no limits to deflections specified for highway structures. There are, however, limits to the deflection of the steelwork supporting the concrete during construction [7].

$$\delta = \frac{(L + 40)}{2000} \tag{4.11}$$

For railway bridges, particularly those on high-speed lines, deflections are more critical and typical limits are outlined in Table 4.2. These limits lead to stiffer deeper structures for bridges carrying railways. A significant proportion of the load of a bridge may be the weight of the structure and its finishes; these deflections are normally precambered out such that at the end of construction the bridge is at its intended level. For a composite bridge the deflections will occur in stages; under the bare steel with wet concrete, the short-term composite and in the long term as creep and shrinkage affect the section. The deflections at these stages should be calculated and set on the steelwork drawing to allow them to be incorporated when fabricated (see Figs 3.5 and 4.8). Generally it is sufficient to give the deflection under only steel

Table 4.2 Deflection limits for railway bridges

Type of movement	Limit
Span/deflection limit at midspan	350 for 1 or 2 adjacent spans 450 for 3 to 5 adjacent spans 1250 for high-speed railways
Twist of track	0.0025 rad on any 3 m length 0.0005 rad for high-speed railway
End rotation at joint	0.005 rad for direct fixing 0.010 rad for ballasted track 0.0035 rad for high-speed railway

(such that these can be measured in the fabrication shop if required), the deflection when concrete is placed and the total precamber required.

The designer and builder should accept that there will be some inaccuracy in the prediction of deflection and precamber; factors that affect the accuracy are numerous. The steelwork will be fabricated using welding; some weld shrinkage will occur and the fabricator will make some allowance for this. Standards [56] allow for inaccuracies in fabrication; a typical allowance for such inaccuracies would be the section length divided by 1000. For a section of girder near the transport limit of 27.4 m this is 25–30 mm. The thickness of the concrete slab is specified but again there may be some variation, which will cause the deflection to vary slightly. The sequence of placing the concrete may also affect deflections. A bridge constructed using a span–span–pier technique (Fig. 4.9(b)) will have a different deflection to that where the slab is cast in one pour (Fig. 4.9(a)).

For a continuous bridge the concrete at supports is unlikely to be cracked straight after construction, leading to some difference with the idealised cracked section of design. The thickness of the surfacing will also vary slightly. The designer should consider these factors and allow some additional precamber if necessary; an allowance

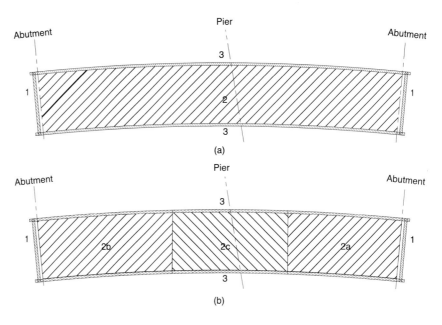

Figure 4.9 Slab construction sequence for Example 4.1: (a) single pour; the pour must be complete before the concrete starts to set to avoid movement and weakening of the steel–concrete interface; (b) span–span–pier casting sequence.

of span/1000 or one third of the live load deflection of a highway bridge has been found to be a reasonable allowance. If the bridge is relatively flat this allowance will also ensure that there is a slight upward precamber; horizontal flanges can look as though they are sagging due to an optical illusion. The designer should also consider any additional long-term deflections; if the bridge is at its correct level during construction then some sag below this will occur in the longer term. If precamber for this is added then the bridge will not be at the correct level at the end of construction.

For the example of Walsall Road the midspan deflections are tabulated below. Given that fully cracked section properties were used over the pier, the permanent plus $L/1000$ or permanent plus one third of live are likely to be overestimates, and a precamber of 93 mm is used.

The deflection for concrete is 54 mm (Fig. 4.7), the limit for Equation (4.7) is 39 mm, so consideration needs to be given to the construction sequence, which must be specified on the drawings (as Fig. 4.9). If so the deflection limit can be increased to:

$$\delta = \frac{L}{300} \tag{4.12}$$

The example is comfortably within this limit.

Natural frequency

The natural frequency (f_n) of a structure is a function of its mass (m) and stiffness (K):

$$f_n = \frac{1}{2\pi}\left(\frac{K}{m}\right)^{0.5} \tag{4.13a}$$

The stiffness is the ratio of load to deflection; hence the first frequency of a bridge can be estimated by knowing only its deflection under permanent loads.

$$f_n = \frac{1}{2\pi}\left(\frac{9.81}{\delta}\right)^{0.5} \tag{4.13b}$$

The frequencies of various bridge structures are shown in Fig. 4.10. Highway structures tend to be a little lower in frequency than rail bridges for a given

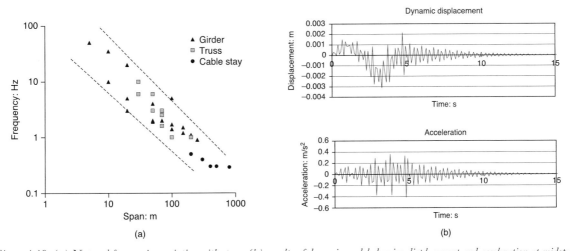

Figure 4.10 (a) Natural frequencies variation with span; (b) results of dynamic model showing displacement and acceleration at midspan.

span. For highway bridges on motorways or major roads, vibration is not usually an issue, although movement under traffic loading may be noticed. Vibration of highway structures is normally only an issue where it is of large amplitude as on long-span structures (see Chapter 10), or if pedestrians use the structure as well as traffic. It is desirable to avoid vibration that may be near resonant frequencies or that people will notice. The natural frequency of a typical vehicle is in the 1 to 3 Hz range and simple dynamic checks should be undertaken if the structure is in this range. A 20 tonne vehicle driving over the structure has been used as a forcing function [30].

For the Walsall Road Bridge the deflection under dead load from the precamber table on Fig. 4.7 is $19 + 54 = 73$ mm. This is calculated using the non-composite section properties; for the calculation of natural frequencies the deflection on the short-term composite is required. The short-term composite second moment of area is approximately 3.8 times greater than the non-composite, and so the deflections will be 3.8 times smaller, the deflection from the surfacing and other permanent loads (6 mm) will also be added. Thus $\delta = 6 + (73/3.8) = 25$ mm. Using Equation (4.13):

$$f_n = 0.16 \times \left(\frac{9.81}{\delta} \right)^{0.5} = 0.16 \times \left(\frac{9.81}{0.025} \right)^{0.5} = 3.6 \, \text{Hz}$$

A 20 tonne vehicle travelling at 50 kph (13.9 m/s) is assumed to be travelling over the structure. The results of the analysis are given in Fig. 4.10(b). The maximum amplitude and accelerations are plotted.

The limits of acceleration or deflection are subjective, for the 20 tonne vehicle a limit of the combined acceleration and displacement [30] is:

$$\Delta a = 0.003 \ (\text{in m}^2/\text{s}^2 \text{ units}) \tag{4.14}$$

where Δ is the dynamic deflection and a the acceleration. For pedestrian-only bridges the acceleration would be limited to [32]:

$$a = 0.5(f_n)^{0.5} \tag{4.15}$$

For Example 4.1 the maximum acceleration is $0.4 \, \text{m/s}^2$ and the amplitude is 0.003 mm. Using Equations (4.14) and (4.15) the calculated values are well below the limiting values and so the bridge should not feel 'lively' to pedestrians when vehicles are passing.

For railway bridges vibration and dynamics are a more significant issue, the higher speeds of trains causing more dynamic loading of the structure.

Through-girder bridges

Through-girders or U-frame bridges have the deck at lower flange level between main girders (Fig. 4.11(b)). The main reason for this is to limit the structural depth below the rail or road level. The transverse beams determine the structural depth below the rail. The width of a through-girder bridge would typically be limited to a single-carriageway road (two lanes with footway) or twin-track railway. More recently [15] a U deck has been developed for smaller spans, the girders and deck are a single fabrication (Fig. 4.11(a)). The through-girder bridge layout is not ideal for steel–concrete bridges as the concrete deck is predominantly in the tension zone and the top compression flange is unrestrained for most of the bridge length.

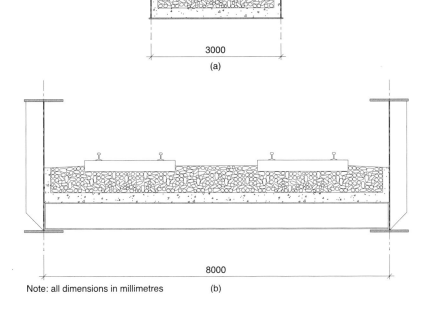

3000

(a)

8000

Note: all dimensions in millimetres   (b)

Figure 4.11 (a) U deck;
(b) through-girder deck.

The transverse deck beams and transverse web stiffeners are used as a U-frame restraint (Fig. 4.12) to partially restrain the flanges. The effective length of the flange for buckling depends on the stiffness of the U-frame:

$$L_e = \pi k (EI_u L_u \delta)^{0.25} \qquad (4.16)$$

If $L_e < L_u$ the U-frame is stiff and flange buckling will occur only between frames. If $L_e > L_u$ the U-frames only partially restrain the flange. The coefficient k is a factor that depends on the support stiffness and whether the load is applied to the tensile or compression flanges. For a rigid end support $k = 0.8$, for flexible end supports $k = 1$. If the load is applied to the compression flange the factor k should be increased by 20%. For stability the design will tend towards the use of stiffer U-frames. However, the stiffer frames will lead to larger live load moments at the girder to cross-beam connection. Through-girder bridges carrying railways

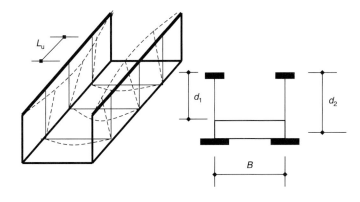

Figure 4.12 U-frame
action.

have experienced fatigue problems at this location; for fatigue the design will tend towards the use of more flexible connections. A key part of the bridge design will be to resolve these conflicting requirements of stability and fatigue.

The U-frame deflection δ is calculated by using the formula:

$$\delta = \frac{d_1^3}{3EI_1} + \frac{Bd_2^2}{2EI_2} + f_j d_2^2 \tag{4.17}$$

The first term estimates the deflection of the web stiffening, the second term the deflection caused by the cross-beam rotation and the third term the deflection from the joint rotation; d_1, d_2 and B are defined in Fig. 4.12; f_j is the joint flexibility which varies from 0.01 radian/MN m for a relatively rigid welded and stiffened connection to 0.05 radian/MN m for an unstiffened bolted joint.

Example 4.2

This is the first of the railway bridge examples; a three-span through-girder structure carries a light rail (metro) system across highly skewed main line tracks. In order to avoid the substantial construction issues associated with large-skew abutments adjacent to the railway, an extra span is added at each side, resulting in a continuous structure with three spans. The span lengths are 30–31 m, 42–43 m and 32–33 m. A benefit of this arrangement is that structurally the girders and cross-beams have no significant skew; the arrangement is also visually more open. The steel girders vary from 2.5 to 3.5 m deep. Cross-beams at 3 m centres span the 9 to 15 m between girders and are 500 to 700 mm deep. A grade 38/45 concrete slab, 250 mm thick spans between cross-beams and supports track plinths (no ballast is used on the bridge).

Loads

The loads on a railway bridge are generally heavier than highway loads. British Standards [32] outline two basic loads, RL and RU; other locomotive (SWL) or high-speed train (UIC) loading may be considered in special circumstances. Load RL approximates light rail or metro systems where mainline rolling stock does not operate. It is primarily applied as a uniformly distributed load of 50 kN/m. Load RU approximates current passenger and heavier goods trains operating on the mainline railway, and is a series of point loads and a heavier uniform load (Fig. 4.13). For this structure RL is appropriate. The rail load is applied to the structure with a partial load factor and a dynamic amplification factor. The dynamic

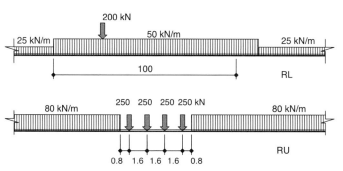

Figure 4.13 RL and RU railway loads.

Note: all dimensions in metres

Table 4.3 Dynamic magnification of loads for railway bridges

Load type	Influence length	Moment	Shear
RU	0 to 3.6 m	2.0	1.67
	3.6 to 67 m	1.0 to 2.0	1.0 to 1.67
	Over 67 m	1.0	1.0
RL	All lengths	1.2 with ballast	1.2 with ballast
		1.4 direct fixing	1.4 direct fixing

amplification factor depends upon the load type and structure length, and is initially estimated from Table 4.3 [32]. To limit weight and dynamic issues some railways are now using track slab technology.

Analysis

Analysis of the structure can again be carried out using a grillage; however, the deflection of the transverse beams will induce lateral loads in the top flange of the main girders that cannot be obtained directly from the grillage. The moments in the flange can be estimated using simplified equations [2] or a beam on elastic foundation analysis. Alternatively a three-dimensional frame analysis could be used to determine coexistent moments in both the vertical and horizontal directions. The beam on elastic foundation model is a useful analysis method for many structural forms including box girders (see Chapter 7). For this example the transverse moments are determined using this method. The results of the various analyses are summarised in Fig. 4.14.

Shear lag

For the multi-beam bridge examples considered so far with spans of 30 to 40 m and beam spacing of 3 to 3.6 m ($b/L = 0.1$) we have assumed plane sections remain

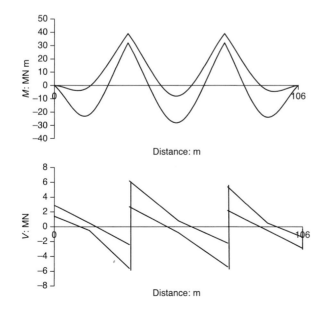

Figure 4.14 Results of the Metro Flyover structural analysis.

61

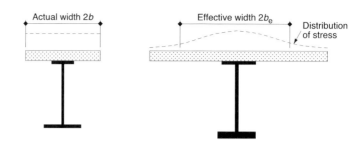

Figure 4.15 Variation in stress across a flange due to shear lag.

plane and that the full width between beams can be utilised. For twin girder bridges where the width is relatively large these assumptions may not be correct. The slab has some in-plane flexibility and when loaded a small displacement occurs, the wider the slab the more flexible it is and the larger the shear displacement or lag. The shear lag causes a variation in longitudinal stress across the section with the stress dropping as the distance from the beam web increases (Fig. 4.15).

For steel structures, allowance for the shear lag is commonly made by using an effective width factor. The factor varies with the width-to-span ratio (b/L), the type of span (simply supported or continuous), the location on the span and the type of load applied. Point loads cause more shear lag than uniform loads. The effective width factors are tabulated in codes [2]. For concrete structures a simpler approach is taken and the effective width is taken to be the width of the beam plus one tenth of the span [3, 7]. For steel and concrete composite structures where the shear lag generally affects the concrete slab, the simpler approach seems logical and the latest composite codes [9] use a similar approach

$$b_{\text{eff}} = b_0 + \frac{L_c}{8} \tag{4.18}$$

where b_0 is the width across the connectors on the beam and L_c is the distance between points of contraflexure.

For calculation of section properties at the serviceability limit state and for fatigue, shear lag should be taken into account. At the ultimate limit state, shear lag can be neglected; however, it is often taken into account as this saves calculation of more section properties. Slab stresses are rarely critical so the degree of conservatism is small. For the Metro Flyover the slab is not continuous beyond the beam and stops at the web, $b_0 = 0$. The span between points of contraflexure is 30 m (0.7L), and at the pier the distance between points of contraflexure is 11 m. Using Equation (4.18):

$$b_{\text{eff}} = 0 + \frac{30}{8} = 3.75 \, \text{m} \qquad \text{at midspan}$$

$$b_{\text{eff}} = 0 + \frac{11}{8} = 1.4 \, \text{m} \qquad \text{at the pier}$$

It can be seen that the effective width at the support is significantly smaller than at midspan. Both widths are significantly smaller than the half-width between girders (6.5 m). Section properties for the midspan and pier-girder sections together with the cross-beams are summarised in Appendix C.

Having ascertained the section properties and the effect of shear lag, the investigation of the U-frame behaviour is continued. For Example 4.2 using Equation

(4.14) and assuming a flexible beam to girder connection:

$$\delta = \frac{d_1^3}{3EI_1} + \frac{Bd_2^2}{2EI_2} + f_j d_2^2$$

$$= \frac{2.6^3}{3 \times 210\,000 \times 0.00032} + \frac{13 \times 3^2}{2 \times 210\,000 \times 0.01} + 3^2 \times 0.05$$

The deflection $\delta = 0.09 + 0.03 + 0.45 = 0.57\,\text{m}$, the flexibility of the stiffener and the joint contributing most significantly to this deflection. To stiffen the U-frame the stiffener could be formed using a T section, stiffening this element by a factor of almost 3, and reducing the U-frame deflection to 0.51 m. Alternatively, the joint at the girder to cross-beam connection could be stiffened, reducing the deflection to 0.15 m. Using Equation (4.13) to estimate the effective length with a flexible frame: $L_e = \pi k (EI_u L_u \delta)^{0.25} = \pi \times 1 (210\,000 \times 0.0019 \times 3 \times 0.51)^{0.25} = 15.6\,\text{m}$, approximately five frames long and significantly shorter than the span.

Using Equation (4.4a), the slenderness parameter (λ) is determined. Since the steel and slab weight are a significant proportion of the load the properties used may be those of the beam (ignoring the slab):

$$\lambda = k_1 k_2 k_3 \frac{L_e}{r} = 1.2 \frac{L_e}{r}$$

Approximately for this example, $\lambda = 108$.

Once the slab is cast the structure is significantly more rigid transversely. Overall lateral torsional buckling will not occur, the buckling will involve only the top flange. The slenderness parameter for this is lower than 108; however, the more conservative figure will be used. From Fig. 2.6 the limiting moment is 40% of the full capacity.

$$M_D = 0.95 Z f_c = 0.95 \times 0.176 \times 0.4 \times 335 = 22.4\,\text{MN}\,\text{m}$$

This is greater than the applied moments of 20.1 MN m. For the top flange, transverse bending, $Z = 0.051\,\text{m}^3$.

$$M_{DT} = 0.95 Z f_c = 0.95 \times 0.051 \times 0.5 \times 335 = 8.1\,\text{MN}\,\text{m}$$

This is again greater than the applied moments of 5 MN m. Both the primary and transverse bending will induce compressions into the flange and the interaction between primary and transverse bending is checked:

$$\frac{M}{M_D} + \frac{M_T}{M_{DT}} < 1 \tag{4.19}$$

$$\frac{20.1}{22.4} + \frac{0.5}{8.1} = 0.90 + 0.06 = 0.96 < 1 \text{ and so is satisfactory}$$

Fatigue

A steel component subject to a series of load cycles may fail in shear if the stress range during the cycles is large. Figure 4.16(a) shows an indication of the critical stress range for a shear stud connector and a channel or bar connector with various loading cycles. Fatigue assessment for both rail and highway loads can be carried out in two ways: a simplified procedure that estimates a limiting stress range, or a more complex damage calculation method. For design, the simplified method is preferred. It may be a little more conservative in some circumstances than the damage calculation method. Given the significant changes in the loading for both highway and railway structures over the last 50 years, some conservatism is justified. The tonnage of freight carried by rail has plunged significantly in this period, while that carried by road has risen (see Chapter 10).

The simplified procedure first involves determining the stress range in the connector, or other component being assessed. The stress is assessed at unfactored or working loads. For railway bridges the stress range is calculated using RL or RU loading as appropriate. For road bridges, the stress range is calculated using a 32 tonne vehicle. Second, the allowable fatigue stress (f_A) on the component is determined and compared with the actual stress:

$$f_A = k f_r \tag{4.20}$$

where f_r is the limiting stress at a given number of cycles. For railways 10^7 cycles is usually assumed [5] with 3×10^8 cycles appropriate for structures on busy motorways. The limiting stress is very dependent on the number of cycles; from Fig. 4.16(a) there is a difference of a factor of 2 between the limiting stress at 10^7 and 3×10^8 cycles ($f_r = 40\,\text{N/mm}^2$ and $18\,\text{N/mm}^2$, respectively). The coefficient k is a factor that varies with design life, the number of tracks or highway lanes carried and the bridge length. Both k and f_r need to be determined from standards [5] for the particular structure being designed. Figure 4.16(b) outlines typical limiting values of f_A for a dual-carriageway highway structure and a lightly and heavily trafficked rail bridge, for shear stud connectors.

For Example 4.2, based on static longitudinal shear requirements, six connectors per linear metre are required near the quarter point. At this location there is both a longitudinal shear and a direct tension effect from the transverse cross-beam, and fatigue of these connectors may be a significant design issue. The shear on the main girder at this location from the RL load is $1210\,\text{kN/m}$. From Appendix C, $Q_1/V = 0.156$, so $Q_1 = 189\,\text{kN/m}$. The tension in the slab caused by transverse bending is $180\,\text{kN}$; this force is assumed to be concentrated over a 1 metre-wide strip near the stiffeners. Using Equation (1.18) the connectors will be subject to the following combined shear and tension load:

$$Q_{1_{max}} = (Q_1^2 + 0.33 T^2)^{0.5} = (189^2 + 0.33 \times 180^2)^{0.5} = 215\,\text{kN}$$

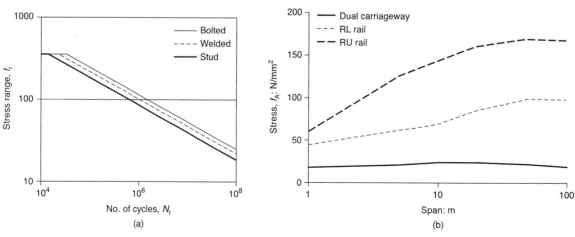

Figure 4.16 (a) *Typical stress ranges with cycles to failure;* (b) *variation in allowable fatigue stress for various spans and load types for shear connectors.*

The load in one connector, $P = Q_1/No = 215/6 = 36\,\text{kN}$. The stress in a connector is calculated from:

$$f = \frac{P}{P_\text{u}425} \qquad (4.21)$$

The stress $f = (36/108) \times 425 = 141\,\text{N/mm}^2$. From Fig. 4.16 the allowable stress for a 40 m span is $99\,\text{N/mm}^2$. Because $f > f_\text{r}$, fatigue is the critical design case and the number of connectors will be increased to 10 per metre. Since there is a tension in the slab and connector, additional reinforcement will be placed to carry the tension and to prevent splitting of the slab. Tests indicate that splitting of the slab may occur prior to connector failure where connectors are placed at the end of a slab.

5

Viaducts

...the most efficient form of structure is a two girder system...

Introduction

Viaducts are major linear structures with multiple supports; they utilise the same beam and slab layout as outlined in previous chapters. Owing to the normally repetitive spans, it is more usual to look for the optimum span and structural layout. The construction method used can often have an influence on the bridge form, and again some repeatable process is likely to be the most economic. Viaducts have been built in many forms; experience has shown that simply supported spans or forms with large numbers of joints and bearings can create maintenance issues [59]. Nowadays, continuous forms minimising bearings and joints are encouraged [40].

Concept design

The first stage in a viaduct design is to determine the most economic material, a steel–concrete composite structure may not necessarily be the best form. Cost differences between well-designed all-concrete structures (or even an all-steel structure for longer spans) and steel–concrete composite structures are relatively small. Often the founding conditions or the nature of the obstacle crossed influences the choice. Where shorter spans can be used and where founding conditions are good a concrete scheme may be better. For larger spans or where the lighter deck of a steel–concrete composite structure leads to significant savings in pile numbers then it is the logical choice. The obstacle crossed may also dictate certain construction methods, that is launching or cantilevering; the experience of the contractor building the structure will heavily influence this form. The perceived maintenance regime will also affect the choice: many clients require the additional costs of the future repainting of steel elements to be taken into account when considering options [60].

Assuming a steel–concrete composite structure is chosen its form must be ascertained. Up to now the examples used in previous chapters have generally been multi-beam forms (Fig. 5.1(a)). Sometimes in order to try to minimise the number of piers and bearings the piers may be inset (Fig. 5.1(b)); this feature will lead to complex details at diaphragms, and increase the steelwork tonnage, particularly where piers are at a skew. The intersection of main girders and cross-diaphragms will have issues with lamella tearing and may require a higher quality steel or additional testing. The multi-beam form is relatively simple on straight bridges but can become complex for curved structures. The multi-beam type is also not the most efficient form particularly for wider structures, as each individual beam has to be designed to carry its share of a heavy abnormal vehicle, leading to more steel in webs and flanges.

The most efficient form of structure for a viaduct is a two-girder system. A number of variations on this form are viable, the most common twin girder form being the ladder beam (Fig. 5.1(c)), which has two main longitudinal girders with transverse cross-beams at a longitudinal spacing of 3 to 4 m. Permanent formwork is often used to form the slab between cross-beams. The forming of the edge cantilevers is often the most difficult feature and so the cantilever overhang is generally limited to about 1.5 m. For long spans with deep girders, short cantilevers can be visually distracting. A variation on the ladder format is to extend the cross-beams beyond the girders to form steel cantilevers (Fig. 5.1(d)). This allows the whole deck to be constructed using the permanent formwork system but will increase steelwork tonnage and can significantly complicate fabrication at the girder to cross-beam connection.

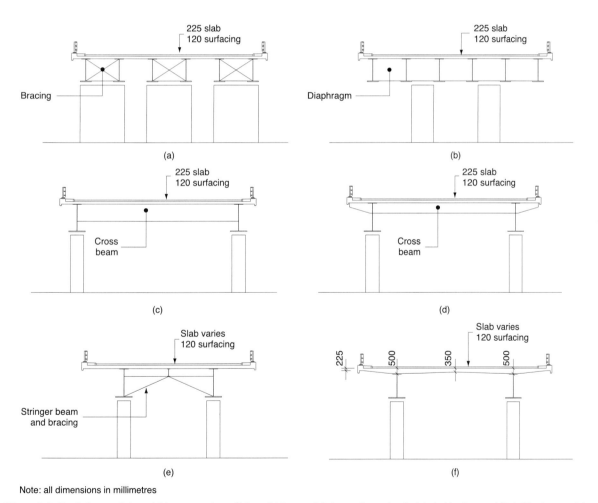

Note: all dimensions in millimetres

Figure 5.1 Deck forms: (a) multi-beam on piers; (b) multi-beam with integral crosshead; (c) ladder beam; (d) ladder beam with cantilevers; (e) twin beam with stringer beam; (f) twin girder.

Another variation on the twin-girder form is the stringer-beam system (Fig. 5.1(e)), this reduces the slab span but can increase fabrication complexity because of the truss system supporting the stringer. The final variation considered is the plain-girder system (Fig. 5.1(f)) where the primary steelwork is limited to the main girders with minimum bracing to provide stability. The slab is profiled and designed to carry all the loads spanning transversely. The slab thickness for this form is greater than the others and so may lead to additional steel tonnage (but not fabrication costs). A formwork gantry system would normally be used to form the slab.

Example 5.1

It is not always easy to decide upon the best form of structure or its materials. The preferences of the client, designer and builder can drive a broad spectrum of solutions. It is often simpler to look at a proposed design and assess it against the key criteria and modify it or propose an alternative. The first example in this chapter is the Doncaster North Viaduct. In the first part of the example the thought process that led to the design is explored.

The need for a road bypassing the centre of Doncaster has been required for some time; the form of the viaduct had been chosen as a post-tensioned, twin-concrete-box design on spans of approximately 40 m (Fig. 5.2(a)). This form is economic and a number of structures of this type have been constructed recently; the author has designed a similar concrete structure as an alternative to the proposed steel–concrete composite at another location [61]. A prestressed concrete

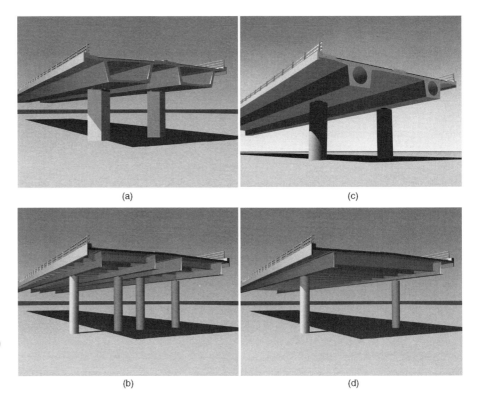

Figure 5.2 Alternative forms considered for the Doncaster North Viaduct: (a) prestressed concrete boxes; (b) four-girder steel–concrete composite; (c) a concrete twin rib; (d) steel–concrete composite ladder beam (© Benaim).

box can be challenging to construct for those without previous experience and two alternatives were considered, a steel–concrete composite, twin-ladder-beam design and an insitu-concrete twin rib (Figs 5.2(b) and (c)). The original design used spans of 40–45 m primarily as this was the length required to span the river, canals and railways along the line. A study of the most economic spans indicates a relatively flat curve over the 30 to 50 m range and tends to show the concrete and steel–concrete composite options with very similar costs. The steel design offered offsite prefabrication of the deck and could be launched or lifted into place; the concrete design with its minimum materials, reduced number of structural elements and simpler foundations was, however, more difficult to build over the railways. A fourth scheme was developed that took the minimal elements of the concrete rib and applied it to a steel–concrete composite scheme cutting down the main longitudinal girders from four to two, and used the pier on single pile foundations (Fig. 5.2(d)). This design was chosen as the preferred scheme and worked up in detail.

The viaduct superstructure consists of a steel–concrete composite ladder-beam form. The twin main girders consist of 2.4 m-deep, constant-depth, steel girders. The top and bottom flanges were a constant width with the thickness varying to suit the bending moments; visually this is far neater than a varying-width flange. Transverse cross-beams at 3.8 m centres span the 16 m between girders. A 225 mm-thick, grade 37/45-concrete slab spans across the beams. The viaduct runs for 620 m starting near the Yorkshire Canal, it runs north across marshy industrial land, railway sidings and then curves across the East Coast Main Line Railway (ECML) and other lines that converge at Doncaster, finally crossing the river Don and ending at an embankment on the flood plain. Flexible pier foundations were used to provide an articulation with multiple fixed bearings.

Articulation

For short bridges it is possible to design bridges without bearings or joints. For lengths up to approximately 80 m it is possible to use integral bridges without joints (see Chapter 3). For bridges over 80 m long, both joints and bearings are usually used. The bearings allow the structure to move and accommodate the strains imposed by temperature, creep or shrinkage, without inducing significant stresses. Expansion joints are used at the ends of the bridge to accommodate this movement and to allow vehicles or trains to pass over the gap. It is possible to design long bridges without bearings if the piers can be made flexible enough, for example the 1600 m-long Stonecutters Bridge (see Chapter 10) has 70 m-high piers fully built into the deck and no bearings.

Expansion joints and bearings have a lower design life than the main structural elements of the bridge and have to be replaced a number of times over the life of the structure. The number of joints and bearings should normally be reduced to a minimum. Proprietary expansion joints and bearings for highway structures can accommodate movements of ±1000 mm, meaning continuous lengths of viaduct of 2 km are possible [61]. Railway joints of a similar size to highway joints have been used but tend to be complicated and expensive. A maximum length of 1200 m has been used for recent high-speed lines [62]. A lower length of approximately 80 m is the limit for unjointed continuously welded track. To avoid track joints, railway viaducts are often formed from a series of short structures each less than 80 m long (Fig. 5.3(b)).

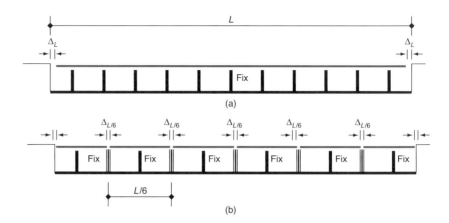

Figure 5.3 Articulation arrangements for railway viaducts: (a) continuous structure; (b) multi-structure form.

Traditionally bearings have been laid out with a fixed point near the centre of the structure and guided or free bearings beyond this, all pointing towards this bearing. This arrangement tends to require a large bearing at the fixed point and often a larger pier and foundation to resist the applied loads. For curved alignments the movements at the expansion joints using this articulation also involve some transverse movements of the joints; this can lead to joint tearing in some situations. An alternative, 'flexible fixity', articulation is to make the piers flexible and fix (or make integral) several bents, spreading the load to a number of piers. Beyond the fixed section the bearings are aligned parallel to the bridge girders. For a curved alignment this means that at the expansion joint movements are only along the structure; however, there are additional transverse forces on the piers.

The layout of the fixed bearings for the Doncaster example is shown super-imposed onto the aerial view of Fig. 5.4. For this viaduct with fifteen spans of

Figure 5.4 Aerial view of the Doncaster North Viaduct showing fixed bearings and joint locations (© Amec Ltd – Construction Services).

approximately 45 m, the permanent reaction (R) at a pier is 10 MN. Assuming a maximum coefficient of friction (μ) of 4% and a minimum of 2% [61], the longitudinal force on any typical pier will be:

$$F = \mu R \tag{5.1}$$

$$F_{max} = 0.04 \times 10 = 0.4 \, \text{MN}, \quad F_{min} = 0.02 \times 10 = 0.2 \, \text{MN}$$

With a single fixed-bearing line near the centre of the bridge the bearing will be required to carry the out-of-balance friction load:

$$F_F = \sum F_{max(F-i)} - \sum F_{min(F+i)} \tag{5.2a}$$

For the example $\sum F_{max} = 7 \times 0.4 = 2.8 \, \text{MN}$, $\sum F_{min} = 7 \times 0.2 = 1.4 \, \text{MN}$. The friction load $F_F = 2.8 - 1.4 = 1.4 \, \text{MN}$, significantly larger than the typical force on other bearings, and larger than the live load braking and traction loads defined in standards [32].

If now we consider the central four pier lines to be flexibly fixed, there are less piers each side of the fixed zone. Then $\sum F_{max} = 6 \times 0.4 = 2.4 \, \text{MN}$, $\sum F_{min} = 6 \times 0.2 = 1.2 \, \text{MN}$. At the fixed section the load is spread between increased number (No) of fixed pier lines:

$$F_F = \left\{ \sum F_{max(F-i)} - \sum F_{min(F+i)} \right\} \frac{1}{No} \tag{5.2b}$$

For this articulation on the viaduct $F_F = (2.4 - 1.2)/4 = 0.3 \, \text{MN}$, which is similar to the forces on other piers and significantly less than that of the single fixed pier articulation.

For the steelwork of a major viaduct like that of Example 5.1 the design is optimised, a small saving to the weight of an element used many times can be significant. The optimisation will use methods outlined in previous chapters but with a number of iterations to develop the best design. Two areas are considered in more detail: the effect of construction methods and deck slab design.

Construction methods

The method of construction will affect the girder size, the need for wide, stable girders in crane-erected sections or thicker webs where girders are launched. For elements lifted into place independently the first elements should be sized to be stable without the subsequent elements. For the Doncaster example the viaduct beyond the railway and river was built element by element with the main girders being placed prior to lifting-in of the numerous cross-beams. Over part of its length the viaduct is curved, and the curvature means there is a tendency for the girder to twist sideways. To prevent the twisting, temporary counterweights are used until the cross-beams are placed. The flanges are sized to give a stable structure for the girders plus counterweights, without the stabilising cross-beams. The majority of small-to-medium-span composite bridges are built using cranes, and for bridges with good access and space to manoeuvre a crane this method is simple and very economical. Components must be of a size that is transportable to the site; in the UK this generally means a maximum length of 27.4 m and a width or height of 4.3 m.

Over the railway and rivers at Doncaster the bridge was placed using a single 'big lift' for the steelwork and permanent formwork (Fig. 5.5). During the lifting, the

Figure 5.5 A big lift of a complete span over the railway at Doncaster (© Amec Ltd – Construction Services).

support locations for the steel will be different to those of the final situation and may require checking. The buckling mode in this case involves two girders and the transverse beams. In this structure the stability of the girders during concreting will be the more critical condition. The loading at this stage involves the steelwork, permanent formwork, concrete and an allowance for live loading during concreting, this is assumed to be $2.5\,\mathrm{kN/m^2}$ and allows for minor plant, men and some mounding of the concrete as it is placed before being spread [63]. The total design load at ultimate limit state during the bridge deck construction is $10.5\,\mathrm{MN}$. The moment on the deck (resisted by two girders) is estimated as:

$$M = 0.072GL = 0.072 \times 10.5 \times 42.5 = 32.1\,\mathrm{MN\,m}$$

If the stiffness of the transverse beams is such that they form a stiff frame (such that $L_e < L_u$ using Equation (4.13)), then for a single girder the buckling length will be limited to the girder spacing. This is a short length, therefore the buckling of a single beam is unlikely to be critical; however, the transverse beams must be sufficiently deep and stiff to achieve this. For the pair of girders, instability may occur with the lateral movement and rotation of the frame, the elastic critical moment (M_{cr}) being dependent on the torsional M_{cT} and warping M_{cW} strengths of the system [64]:

$$M_{cr} = (M_{cT}^2 + M_{cW}^2)^{0.5} \tag{5.3a}$$

$$M_{cT} = \frac{\pi}{L}(EIGJ)^{0.5} \tag{5.3b}$$

$$M_{cW} = \frac{\pi^2}{L^2}E(IC_w)^{0.5} \tag{5.3c}$$

where G is the shear modulus, J is the torsional constant (see Appendix E) and C_w is the warping constant. The slenderness parameter of the girder system is estimated from the following:

$$\lambda = \left(\frac{\pi^2 EZ}{M_{cr}}\right)^{0.5} \tag{5.4}$$

For most plate girder bridges the torsional resistance (M_{cT}) is small and most of the resistance comes from the warping resistance of the girder pair, $M_{cr} = M_{cW}$. The warping resistance (M_{cW}) is primarily dependent on the distance between girders [65]:

$$C_w = B^2 \frac{D^3 t_w}{24} = \frac{B^2 D^3 t_w}{24} = \frac{16^2 \times 2.4^3 \times 0.02}{24} = 3\,\text{m}^6 \tag{5.5}$$

For the Doncaster beams using Equations (5.3) and (5.4):

$$M_{cW} = \frac{\pi^2}{L^2} E(IC)^{0.5} = \left(\frac{9.9 \times 210\,000}{42.5^2}\right) \times (0.2 \times 3)^{0.5} = 930\,\text{MN m}$$

$$\lambda = \left(\frac{\pi^2 EZ}{M_{cr}}\right)^{0.5} = \left(\frac{9.9 \times 210\,000 \times 2 \times 0.071}{930}\right)^{0.5} = 12$$

From Fig. 2.6, this means the full capacity of the section can be carried. Using Equation (1.8) for the two girders: $M_D = 2 \times 0.95 \times 345 \times 0.071 = 47\,\text{MN m}$, which is greater than the applied 32.1 MN m moment.

An unpropped crane erected method of construction has been used in all the previous examples, it is the most economic method for short and medium spans. Propping of the steelwork can be used instead of bracing to reduce the span and effective length during construction, this also reduces the moment on the non-composite section and allows some reduction to both top and bottom flanges. Propping the steelwork prior to concreting can aid slender or non-compact sections; the majority of the load is immediately carried by the composite section when the props are removed. If a single prop is used, the moment in the steelwork reduces to approximately 25% of the unpropped value, and the top flange can be reduced. As the steel and concrete dead load accounts for about one third of the total design load, which is now largely carried by the composite section with a 30% greater modulus, it may be possible to reduce the bottom flange. Overall, propping may save 5% to 15% of the girder steelwork tonnage. There will be little change to fabrication and painting costs as the number of stiffeners, weld volume and overall painted area will hardly be affected. The cost of the temporary prop towers and their foundations will need to be added to the erection costs, usually exceeding the cost of the saved material. Propping is more often used for larger spans in combination with a launched construction method (see Chapter 8).

Launching of the Doncaster viaduct across the ECML was considered at an early stage but the large lift was felt to be more appropriate for the site situation. A number of viaducts over railways have been launched into position. Launching involves the sequential building and launching of a viaduct from one abutment out over the obstacle; a new section is added to the rear at each stage. A launching nose is normally used to limit the cantilevering stresses in the steelwork. The nose is shaped such that it compensates for the cantilever deflections, gradually bringing the steelwork back to level. A nose length of approximately 70% of the span is normally used for a concrete deck, less if only steelwork is launched.

The moment range in the structure during launch is shown in Fig. 5.6. It can be seen that all sections of the structure need to be designed to cope with a significant moment range. The support points of the structure move as launching progresses and will bear on unstiffened portions of the web; a check on crippling and buckling may be required unless an overthick web is used (see Chapter 8). It is common to launch only the steelwork as this limits the moments and shears in the structure and limits the jacking forces required. The launching of the steel–concrete composite section can, however, limit the work required over the railway or other obstacle and has been successfully carried out [66].

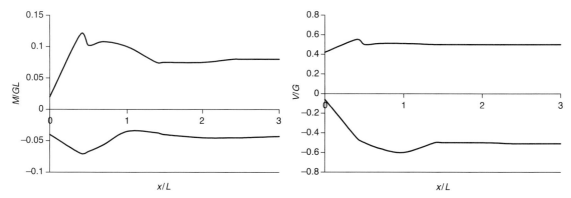

Figure 5.6 Moment and shear envelope for a launched viaduct.

Deck slab

The final aspect of the Doncaster viaduct to be considered in more detail is the issue of the deck slab design and construction. The trend in UK construction practice is for maximum off-site prefabrication with minimised on-site work. The steel elements of a steel–concrete composite viaduct fit with this trend, the insitu-deck slab does not. A fully insitu-deck slab requires formwork and falsework to support the wet concrete, and these have to be removed from between the girders after the slab has cured. A safer way to form the slab is to use permanent formwork allowing some off-site prefabrication and avoiding the need to strip the soffit of formwork. If designed to participate with the deck slab for carrying loads, some economy in deck slab reinforcement may result. The use of a full-depth, precast slab with only insitu stitches gives maximum prefabrication. For the full-depth, precast slab the joint between units and overbeams need to be detailed carefully to achieve continuity (Fig. 5.7).

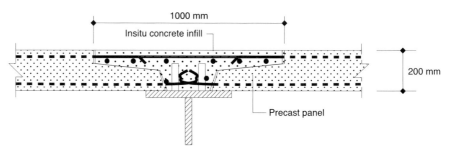

Figure 5.7 Details of insitu stitch on a precast deck slab [67].

Slabs for composite bridges have been traditionally designed as bending elements using fully elastic methods of analysis; it has been known for some time [42, 46] that this leads to a conservative design. For most slab panels, except those of cantilever edge panels, some arching action [68] will exist (Fig. 5.8), and the design issue is how to take this into account. For relatively robust slabs with a span-to-depth ratio of 15 or less and a span below 3.2 m, full arching action may be assumed [47]. A relatively simple internal arch within the slab is assumed to carry the wheel load. The slab capacity is estimated assuming a punching shear failure:

$$V_{arch} = 1.25(2r + d)d(f_{ck})^{0.5}(100\rho)^{0.25} \tag{5.6a}$$

$$\rho = \frac{kf_{ck}h^2}{300d^2} \tag{5.6b}$$

where ρ is an effective reinforcement ratio, r is the radius of the loaded area, d and h are the slab depth and thickness, and k is a coefficient dependent on the span to depth ratio and the strain at which plasticity of the concrete occurs (see Fig. 5.8).

For spans above 3.2 m or those with a span to depth ratio greater than 15, some arching action will occur with bending. In order to estimate the degree of arching action the relative stiffness of the in-plane and bending components needs to be evaluated. The stiffer the in-plane strength of the slab the greater the proportion of load that will be carried as an axial thrust (N) rather than bending (M).

Arching action assumes an internal arch within the slab (Fig. 5.8(a)) with the compression diagonals thrusting against adjacent concrete panels. For internal panels the restraint may be approximated as:

$$K = Eh \tag{5.7}$$

Where the panel is nearer the edge of the bridge the restraint will be less [46, 68], for cantilever edges or panels adjacent to joints no restraint should be assumed. The geometry of the arch will have a significant effect on the arch capacity, it will be dependent on the depth of slab in compression. Typically a $0.2h$ to $0.3h$ compression depth may be assumed, but this should be confirmed during the analysis, particularly if there are significant deflections under load.

The analysis method involves the modelling of a representative beam strip, with the beam, arch and restraints included. The slab of the Doncaster North Viaduct is used as an example of the method. The deck slab is of grade 37/45 concrete, 225 mm thick spanning approximately 3.8 m between transverse cross-beams, it carries loading of 45 units of HB at the ultimate limit state and 25 units at the serviceability

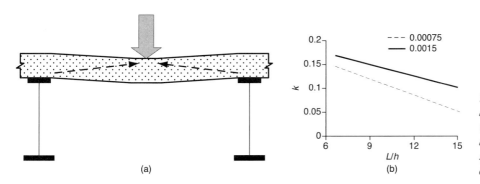

Figure 5.8 (a) Arching action in a slab; (b) coefficient k for arching action in a slab for various span-to-depth ratios and elastic strains.

limit state. The critical design loading is with two axles of the HB bogie slightly offset from the slab midspan (Fig. 5.9).

The use of a conventional strip, influence line or a simplified finite element analysis indicates an ultimate peak hogging moment of 60 kN m/m and a sagging moment of 90 kN m/m. Using Equation (1.5) with $d = 0.75h$ or 170 mm, the ultimate capacity of the slab based on compression of the concrete is:

$$M_u = 0.15bd^2 f_{cu} = 0.15 \times 1000 \times 170^2 \times 45 \times 10^{-6} = 195 \text{ kN m/m}$$

This is two to three times the applied moment and so the depth in compression will be significantly less than half of the assumed slab depth (see Chapter 1). A neutral axis depth of 0.25h at the support and midspan will be assumed for the analysis, giving an arch rise of 0.5h, say 100 mm. The restraint stiffness of the slab at the arch springing is estimated using Equation (5.7), assuming $E_c = 34\,000 \text{ MN/m}^2$ (Table 1.1) and $h = 225$ mm:

$$K = Eh = 34\,000 \times 0.225 = 7600 \text{ MN m}$$

Analysis of the slab strip with the internal arch indicates an ultimate hogging moment of 55 kN m/m and a sagging moment of 78 kN m/m. The arch thrust is 350 kN/m and the midspan deflection of the slab is 12 mm.

The bending capacity of a reinforced concrete section is increased by the application of a moderate, coexistent axial load. An interaction curve for a 225 mm thick slab of grade 45/37 concrete, with various reinforcement percentages, is shown in Fig. 5.10. It can be seen that there is some bending capacity with axial load even for unreinforced slabs. At high compressive stresses the amount of reinforcement makes little difference to the slab capacity.

From the interaction curve for the slab without arching the reinforcement requirement for pure bending is 1300 mm²/m for the 85 kN m/m moment. For the arched slab the reinforcement requirement is 750 mm²/m for the 77 kN m/m moment with 350 kN/m axial load. Figure 5.10 also outlines when the neutral axis reaches the limiting ($d/2$) value in the slab for various reinforcement

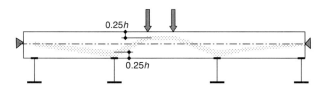

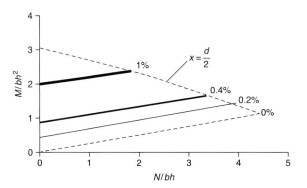

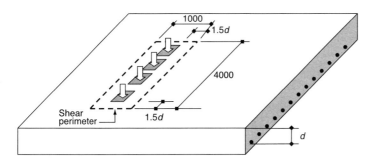

Figure 5.11 Punching shear on the deck slab.

amounts. For the arch the neutral axis depth is estimated to be approximately 40 mm. Reviewing the arch geometry a revised estimate of the arch rise is:

$$r = h - 2x - \delta \qquad (5.8)$$

For this example, this is 133 mm, larger than the rise originally assumed and so the original assumptions are conservative. If the analysis is reiterated using the new arch geometry a reduced reinforcement requirement is obtained due to an increased axial load and reduced moment.

A check on the punching shear capacity of the slab is also carried out. Punching shear is assumed to occur at 1.5d from the wheel load (Fig. 5.11). The ultimate shear force (V) is the vehicle bogie weight of 644 kN and the shear perimeter (L_v) is calculated to be 10 m.

The limiting shear stress (v_c) of a grade 32/40 concrete slab is [3]:

$$v_c = 0.74 k_1 k_2 \left(\frac{100 A_s}{L_v d} \right)^{0.33} \qquad (5.9)$$

where k_1 is a coefficient dependent on the concrete strength and is 1.0 for grade 32/40 concrete or more, and k_2 is a coefficient dependent on the concrete depth. For a 225 mm-thick slab k_2 is 1.22, for a 500 mm slab it is 1.0. The shear resistance of the concrete section is:

$$V_D = L_v d v_c \qquad (5.10)$$

For this example the shear resistance must be at least 645 kN; rearranging Equation (5.10):

$$v_c = \frac{V_D}{L_v d} = \frac{645\,000}{10\,000 \times 170} = 0.38 \, \text{N/mm}^2$$

Rearranging Equation (5.9)

$$\frac{100 A_s}{L_v d} = \left(\frac{v_c}{0.74 k_1 k_2} \right)^3 = 0.075$$

The minimum steel in a slab is typically 0.15% of the concrete area and so punching shear is not critical for this example.

6

Haunches and double composite action

...a double composite structure...providing composite action to both the top and bottom flanges...

Introduction

For short- to medium-span bridges the constant depth girder is an economic solution as fabrication is relatively simple. For longer spans the cracked concrete section over supports becomes less efficient. In order to improve the effectiveness of a continuous structure at intermediate supports the section can be improved by haunching (increasing the girder depth locally), or by the use of a double composite action (the addition of concrete to the lower flange). Both haunching and the use of a double composite section will affect the distribution of forces in the girder and the shear flow at the steel–concrete interfaces.

Haunches

Visually constant depth girders can look heavy, particularly if the deck overhang is small. In general the deck overhang should always be at least the depth of the girder, preferably more. Haunching of the beams allows shallower span-to-depth ratios to be used over the middle section of the span (see Table 4.1).

Consideration of the moment diagram for a bridge or viaduct span (Fig. 6.1), shows that deepening the girder adjacent to supports, where the moments are larger, could form a more structurally efficient arrangement. The deepening of the girder increases the stiffness and draws further moment to the support. The

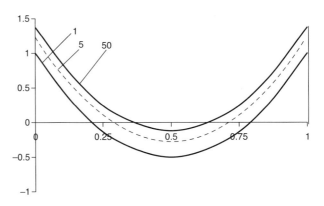

Figure 6.1 Moments on a viaduct span with increasing relative haunch stiffness.

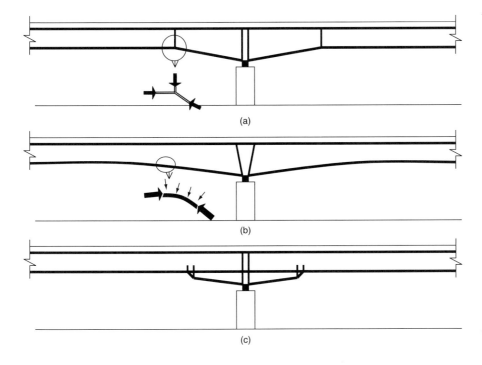

(a)

(b)

(c)

Figure 6.2 Forms of haunching: (a) straight; (b) curved; (c) corbel.

extent of haunching should be carefully considered as drawing too much moment or shear to the support can be detrimental; the web slenderness increases, reducing the limiting shear stress, and the composite section is less efficient with the upper deck in tension and the lower steel flange in compression.

The form of the haunching could be a straight haunch (Fig. 6.2(a)), a gradual curving change in depth (Fig. 6.2(b)) or a corbel type (Fig. 6.2(c)) [75]. The haunching of the steelwork will affect the design process and detailing. For a straight haunch, stiffeners are required to resolve forces where the flange changes direction. For a curved or parabolic haunch, additional in-plane forces are generated in the web with some local transverse bending of the flange, these are usually small enough to avoid the need for stiffening. For both types the flanges are not parallel and so the moment–shear (M–V) interaction criteria (outlined in Chapter 4) may not be valid, as this method assumes some post-buckling strength. Testing has generally been confined to parallel flange beams [54]; where flanges are significantly non-parallel the use of an elastic buckling criteria design method (see Chapter 10) may be more applicable, leading to thicker webs.

Longitudinal shear at changes of section

The longitudinal shear is determined from the rate of change of force in the composite flange adjacent to the interface (Equation (1.14)). For constant-depth girders this was simplified for design purposes to Equation (1.16):

$$Q_1 = \frac{V A_c y}{I} \tag{1.16}$$

For a haunched girder additional forces are generated due to the change in section. For a constant-depth girder element under a uniform moment there is no change in force; for a girder with a changing depth there is a change in force due to the depth change ($M = Fz$, Equation (2.1)). The longitudinal shear due to the change in depth is:

$$Q_1 = M \frac{\mathrm{d}}{\mathrm{d}x} \left(\frac{A_c y}{I_{a-c}} \right) \tag{6.1}$$

The total longitudinal shear on the interface for a haunched girder is the sum of Equations (1.16) and (6.1):

$$Q_1 = \frac{V A_c y}{I_{a-c}} + M \frac{\mathrm{d}}{\mathrm{d}x} \left(\frac{A_c y}{I_{a-c}} \right) \tag{6.2a}$$

Where the flanges are large and the web slender:

$$Q_1 = \frac{V A_c y}{I} \left(\frac{D_2}{D_1} \right) \tag{6.2b}$$

where D_1 and D_2 are the distances between flanges at each end of the section over which the longitudinal shear is being calculated.

Double composite action

A way of further improving the effectiveness of a steel–concrete composite structure at intermediate supports is to use a double composite section. With a double composite structure an additional concrete element is added to the lower (compression) flange providing composite action to both the top and bottom flanges. The lower concrete flange allows the steel elements to be reduced in size owing to both the sharing of load between the steel and concrete sections and the elimination of lateral buckling modes by the introduction of the slab with a high lateral inertia. The design of sections with double composite action follows the basic rules outlined for conventional composite structures, but with both interfaces requiring design.

Example 6.1

This example considers a steel–concrete composite river bridge [69] with a series of 110 m spans. The large 110 m spans are used in the river where ship impact and seismic requirements govern the design. On-shore, shorter spans are used. A number of interesting features are used on the bridge. First, the section is haunched. Second, the lower flanges have a concrete slab between them creating a double composite structure. Third, the structure uses a twin girder arrangement with minimum bracing and cross-girders, the profiled slab forming the primary transverse member of the 19 m wide deck. Finally, a twin pier arrangement is used to increase stiffness further (see Fig. 6.3).

Loads and analysis

The total self-weight of the steelwork and concrete deck are approximately 225 kN/m. Near the pier the lower double composite slab adds a further 80 kN/m. Surfacing and parapets add 75 kN/m and the unfactored live load is approximately 65 kN/m. Analysis of the bridge is by a simple line-beam model. A twin pier arrangement is used to increase the stiffness of the system; this arrangement is often used on concrete bridges but less so on steel–concrete composite structures. For live loads this arrangement doubles the midspan stiffness and is particularly useful for long-span railway structures where deflection limits often govern section sizes.

At the pier the girder is 6.6 m deep. This depth of girder is difficult to fabricate and transport and so the steelwork is formed of two sections, the lower with a small top flange and the upper with a small bottom flange, these smaller flanges being bolted together to form the larger girder section (Fig. 6.3). The depth of the section also affects the behaviour of the web.

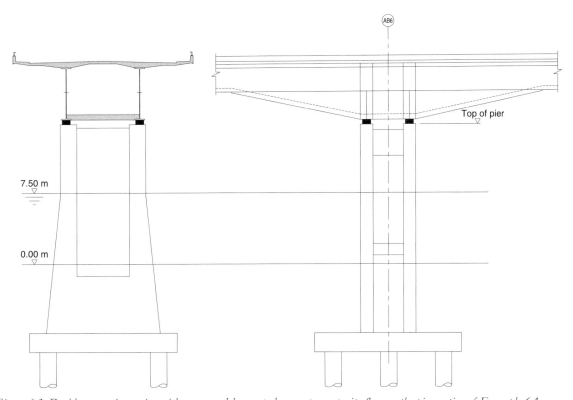

Figure 6.3 Double composite section with upper and lower steel–concrete composite flanges, the pier section of Example 6.1.

Slender webs

For deep girder sections the web slenderness is large and the web is not fully effective; there is a tendency to shed stress from compression areas. This may be approximated by ignoring the part of the web in compression, or by using a reduced effective thickness for the web when calculating stresses. For the reduced effective web method [2], if the depth of web in compression (y_c) is greater than $67t_w$ it is classed as slender and a reduced section should be used for calculating the section properties when calculating bending and axial stresses. The reduced effective web is estimated from:

$$t'_w = 1.425t_w - 0.00625y_c \tag{6.3}$$

Table 6.1 Moment and stress build-up for a double composite section of Example 6.1

Construction stage	M: MN m	Z_t: m^3	Z_b: m^3	f_{at}	f_{ab}	f_{ct}	f_{cb}
Steel cantilever	20	0.55	0.39	-36	51	0	0
Single composite section	40	1.02	0.85	-39	47	0	4
Double composite section	70	1.04	1.37	-67	51	-68	4
Parapets and surfacing added	50	1.04	1.37	-48	37	-49	3
Live loading	60	1.07	2.24	-56	27	-57	5
Total stress: N/mm^2				-256	213	-174	16

If the depth in compression exceeds $228t_w$ then the effective web thickness is zero and all bending is assumed to be resisted by only the flanges.

For this example the depth of the web in compression is initially estimated as 3000 mm, for a 25 mm web the slenderness is:

$$\frac{y_c}{t_w} = \frac{3000}{25} = 120$$

Using Equation (6.3):

$$t'_w = 1.425 t_w - 0.00625 y_c = (1.425 \times 25) - (0.00625 \times 3000) = 17\,\text{mm}$$

For the steel section this reduces the section modulus by about 15% from 0.46 to 0.39 m^3.

The bridge is constructed by cantilevering of the steelwork from the stable twin pier arrangement; the lower concrete flange is placed prior to linking of the cantilevers. The upper deck slab is constructed in two stages using moving formwork gantries from the pier outwards on the continuous girders. The stresses build up stage by stage on the various sections: the non-composite steel section, the section with a composite lower flange and the final section with a composite upper and lower flange. Table 6.1 outlines the moment at each stage of construction and the stresses induced.

The total shear on the section at the serviceability limit state is approximately 24 MN (12 MN per girder). The composite section will carry 85% of the shear. For this girder the Q_l/V ratio is approximately 0.12 for both the upper and lower shear interface. The slope of the haunch is approximately 1 in 5, using Equation (6.2b):

$$Q_l = \frac{V A_c y}{I}\left(\frac{D_2}{D_1}\right)$$

$$= 12 \times 0.85 \times 0.12 \times 1.03 = 1.3\,\text{MN/m}$$

Using Equation (1.17) with 22 mm-diameter connectors (Table 1.5):

$$No = \frac{Q_l}{0.55 P_u} = \frac{1.3}{0.55 \times 0.139} = 17$$

The requirements equate to 4 connectors at 200 mm centres. The layouts of the upper and lower connectors are shown in Fig. 6.4. Longer connectors are used at the outside edges of the upper flange [88], as there is also some uplift on these

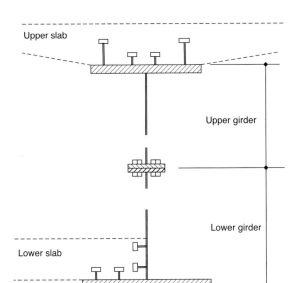

Figure 6.4 Connector details for upper and lower slab of Example 6.1.

connectors from local vehicle loads, particularly adjacent to stiffeners (see Chapter 4 for connector design with additional tensile loads).

Lightweight concrete

Concrete with a reduced density from 14 to 24 kN/m³ may be used in composite structures, particularly in long-span bridges, reducing the overall dead weight of the span. Typically densities of 21 to 23 kN/m³ can be achieved with the use of a lightweight aggregate, to get below 20 kN/m³ the finer aggregate (sand) will also need to be replaced. The strength of concrete is dependent largely on the strengths of the aggregate and generally lightweight concrete can achieve similar strengths to normal concrete. The letters 'LC' adjacent to the grade designate lightweight concrete.

The properties of lightweight concrete are slightly different from normal density concrete. The coefficient of thermal expansion is lower, typically $9 \times 10^{-6}/°C$ rather than 12×10^{-6}, which means that additional longitudinal shear may be induced by changes in temperature. Lightweight concrete is usually less permeable than conventional concrete; creep is usually higher by 20 to 60% depending on the aggregate. At the steel–concrete interface the reduction in the elastic modulus is the most important difference:

$$E_{CL} = \left(\frac{\rho}{24}\right)^2 E_c \tag{6.4}$$

where E_{CL} is the modulus for lightweight concrete of density ρ, compared to a conventional concrete of the same strength. The capacity of the connectors is dependent to some degree on the stiffness of the concrete; Equation (1.19a) is modified for lightweight concrete:

$$P_{UL} = P_u \left(\frac{E_{CL}}{E_c}\right)^{0.5} \tag{6.5}$$

Equations (1.24) and (1.25) governing the ultimate shear strengths of the possible shear planes in the concrete are also affected:

$$Q_1 = 0.7L_s + 0.7A_s f_y \tag{6.6}$$

$$Q_1 = 0.12L_s f_{cu} \tag{6.7a}$$

$$Q_1 = 0.125L_s f_{ck} \tag{6.7b}$$

If lightweight concrete of density $20\,\text{kN/m}^3$ and grade LC 40/50 is used for the deck of Example 6.1, the longitudinal shear will reduce to approximately $1.1\,\text{MN/m}$; however, as the modulus and connector capacity also reduce very similar connector requirements are obtained.

7

Box girders

...a triangular structure...is stiffer and less susceptible to distortion than the conventional ...forms...

Introduction

The first metal-box-girder bridge was built in 1850 by Stephenson over the Menai [70]. Although it was a riveted, wrought-iron structure, it had the basic stiffened plate form used by many bridges in the twentieth century. In the 1970s a series of collapses in the UK, Europe and the Far East highlighted that the box form had developed beyond the knowledge of known behaviour, a phenomenon that possibly occurs with every generation [80]. Extensive research followed the collapse enquiries [54, 81]. The use of large steel-box-girder bridges is still relatively uncommon; their major use is on larger cable-suspended spans (see Chapter 10). Concrete boxes are more tolerant of the shear lag, warping and distortional effects that affect thin steel-plated sections; often diaphragms are not required, making them simpler to construct. Concrete boxes are a common form for 40 to 150 m spans. Composite boxes are an intermediate type, they have an advantage over the all-steel box in avoiding the fatigue-sensitive steel authotropic deck [82]. The design complications of warping, distortion and shear lag still occur. Intermediate diaphragms are used to limit distortion. The fabrication costs of boxes are relatively high compared with plate girders because of access difficulties for welding inside the box. The use of an open-topped steel box with the concrete deck forming the top flange is often preferred to avoid this access issue. Steel–concrete box girders are nowadays most common on moderate spans for aesthetic purposes [16].

Behaviour of boxes

When subject to bending, box girders behave similarly to plate girders; they are subject to buckling, shear lag and local slenderness effects. When subject to a torsional moment, either from eccentric loading or from curvature of the structure, the box forms a much stiffer structure than a plate girder. The resistance to torsion in a box is from a shear flow around the box (Fig. 7.1(a)). The shear flow around the box is uniform:

$$v_w t_w = v_f t_f \tag{7.1}$$

$$T = BD(v_w t_w + v_f t_f) \text{ or } 2A_o vt \tag{7.2}$$

where v_f and v_w are the shear stress in the flange and web, A_o is the area within the box centreline $(B \times D)$ and T is the total torsion moment at the section. The torsional shears in the web will add directly to (or subtract directly from) those shears due to vertical loads (v_v):

$$v = v_v + v_w \tag{7.3}$$

The shear stresses around the section cause a distortion of the box, idealised as a parallelogram (Fig. 7.1(b)) but in real structures involving some transverse bending due to the rotational stiffness at the plate intersections (Fig. 7.1(c)).

Conventional plate girders can carry torsional moments if the torsion is resisted by the warping of a girder pair, one girder moving upwards and the other down. A similar warping occurs in the box section, but because the webs are connected by the flanges the distribution of stresses is slightly modified.

Intermediate diaphragms within the box span will reduce the effects of warping and will restrain distortion, causing some secondary restraining stresses. The relative magnitude of the stresses depends upon the structure geometry. The stresses can be derived from an analysis of the box (using a beam on elastic foundation [83] or a finite element model) or from simplified equations [2], provided certain geometric limits are satisfied. The simplified method will give an indication of the stresses; where the stresses are significant the more complex analysis may be required, or a more efficient structural arrangement sought. At the junction of the web and flange, when subject to an increment of torque T_i the longitudinal stress induced by the restraint to torsional warping is:

$$f_{TW} = \frac{D T_i}{J} \tag{7.4}$$

where J is the torsional constant (see Appendix E). This stress drops rapidly each side of the applied torque at a distance a from the applied torque:

$$f_{TWx} = f_{TW}\, e^{-(2a/B)} \tag{7.5}$$

This means that generally there is little interaction between increments of torque further apart than the box width. The longitudinal distortional warping stress (f_{DW}) can be estimated from:

$$f_{DW} = \frac{k_1 L_D^2 T_{UD}}{BZ} \tag{7.6}$$

where L_D is the diaphragm spacing and T_{UD} is the applied uniformly distributed torque. When βL_D is less than 1.6, k_1 is a constant at 0.22; when βL_D is greater than 1.6, $k_1 = 1/\{1.66(\beta L_D)^2\}$.

$$\beta L_D = (k_2 L_D^4 / EI)^{0.25} \tag{7.7}$$

$$k_2 = \frac{24 D_{YT} k_3}{B^3}$$

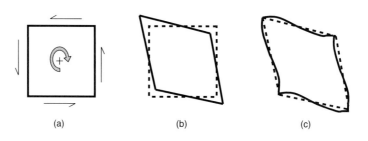

Figure 7.1 Distortion of box: (a) shear flow around section; (b) idealised distorted shape; (c) distortion with transverse bending.

(a)   (b)   (c)

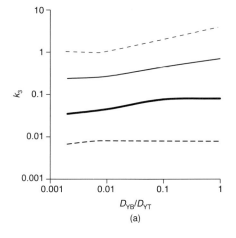

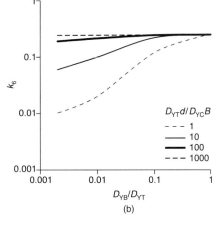

(a)

(b)

Figure 7.2 (a) Coefficient k_3 for longitudinal distortional warping; (b) coefficient k_6 for transverse distortional bending.

where k_3 is obtained from Fig. 7.2(a) and depends on the relative transverse flexural rigidity of the top and bottom flanges (D_{YT} and D_{YB}) and webs (D_{YW}).

The longitudinal torsional and distortional warping stresses will add directly to (or subtract directly from) the bending stresses in the girder:

$$f = f_b + f_{TW} + f_{DW} \qquad (7.8)$$

Transverse bending (f_{DB}) effects from the distortion (Fig. 7.1(c)) can be estimated from:

$$f_{DB} = \frac{k_4 k_5 T_{UD}}{B Z_f} \qquad (7.9)$$

When βL_D is greater than 2.65, $k_4 = 1$; when it is less than 2.65,

$$k_4 = 0.5(\beta L_D)^{3.7}$$

$$k_5 = 0.5 B k_6$$

where k_6 is obtained from Fig. 7.2(b) and again depends on the relative flexural rigidity of the flanges and webs, and Z_f is the transverse flange modulus.

Diaphragms

The behaviour of the box is influenced by the stiffness and spacing of the diaphragms. Typically diaphragms will be at 2 to 4 times the box depth to limit the distortion and warping stresses. The diaphragms should also be stiffer than the box section; a figure of 1500 times that of the box is recommended [2]. For Equations (7.6) and (7.9) to be valid a minimum diaphragm stiffness (K_D), as shown in Fig. 7.3(a), is required, where the stiffness is determined from the application of a series of diagonal loads applied to the corners of the diaphragm (Fig. 7.3(b)).

Three forms of diaphragm are commonly used: the plate diaphragm, braced diaphragm and ring diaphragm. The plate diaphragm is relatively simple to fabricate and has a good stiffness. Access requirements through the box will complicate the diaphragm layout and additional stiffening will be required. The ring diaphragm is even simpler and allows maximum access through the box, but it is more flexible.

Figure 7.3 (a) Minimum diaphragm stiffness requirements; (b) force system to determine the diaphragm stiffness across the diagonal.

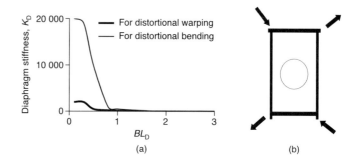

(a)   (b)

The braced diaphragm allows a compromise to be achieved between access and stiffness.

Example 7.1

In the UK, steel boxes are a popular form for railway bridges. The box is more stable than a conventional plate girder, provides a stiffer U-deck and is relatively simple to construct quickly during railway possessions. The first example in this chapter is a box of this type.

The example is a 22 m-span twin box bridge carrying a railway over a road in the Midlands. The form of the structure is a U-frame (see Chapter 4) with the deck between the boxes; the U-frame action only provides limited restraint, as the torsional stiffness of the girder is relatively large. The layout of the steelwork is shown in Fig. 7.4. The box, cross-beams and detailing of the box to beam connection are designed to allow fast assembly during a railway possession, one face of the box being inclined to provide more space and to ease assembly of the deck beams.

Loads and analysis

The bridge carries its self-weight, a layer of ballast and full RU loading. At the ultimate limit state the loading on the girder is approximately 330 kN/m; this load is applied to the edge of the box via the cross-beam connection inducing both a moment and torsion in the box. The total load on the 22 m span $(G + Q)$ is 7.3 MN, giving $M = 20$ MN m, $V = 3.9$ MN, $T_{UD} = 0.15$ MN m/m and $T = 1.6$ MN m. The torsional properties of the box are summarised in Appendix F and the bending properties in Appendix C.

The vertical and torsional shears in the box are calculated:

$$v_{\mathrm{v}} = \frac{V}{dt_{\mathrm{w}}} = \frac{3.9}{1.88 \times (0.025 + 0.015)} = 53 \, \mathrm{N/mm}^2$$

the torsional shear being estimated from a rearrangement of Equation (7.2):

$$v_{\mathrm{T}} = \frac{T}{2A_{\mathrm{o}}t} = \frac{1.6}{2 \times 1.8 \times 0.025} = 18 \, \mathrm{N/mm}^2$$

Using Equation (7.3) the maximum shear stress is estimated:

$$v = v_{\mathrm{v}} + v_{\mathrm{w}} = 53 + 18 = 71 \, \mathrm{N/mm}^2$$

From Fig. 1.6, with $d/t = 80$ and a panel aspect ratio of 2 the limiting shear stress is $0.86v_{\mathrm{y}} = 167\,\mathrm{N/mm^2}$; this is greater than the applied shear and so the web plate is satisfactory.

The bending stress at midspan of the box is calculated:

$$f_{\mathrm{b}} = \frac{M}{Z} = \frac{20}{0.08} = 250\,\mathrm{N/mm^2}$$

The width of the box is small, so T_{i} is similar to T_{UD} for this example. The longitudinal stress due to torsional warping is (Equation (7.4)):

$$f_{\mathrm{TW}} = \frac{DT_{\mathrm{i}}}{J} = \frac{2 \times 0.15}{69} = 0.004\,\mathrm{N/mm^2}$$

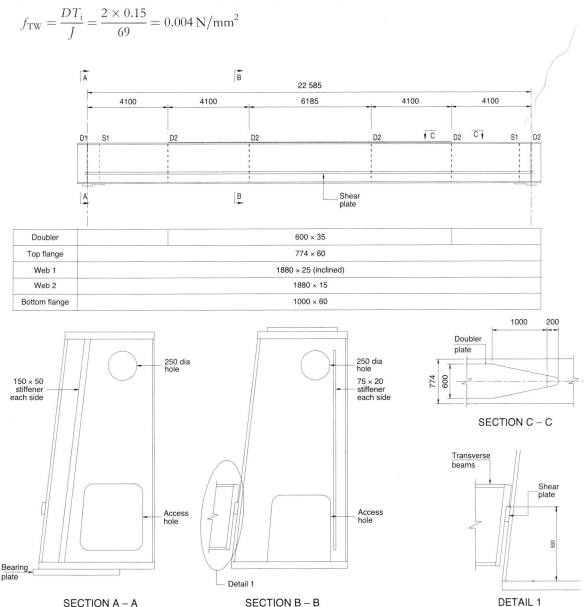

Note: all dimensions in millimetres

Figure 7.4 Typical standard box railway bridge details.

Table 7.1 Longitudinal warping and transverse distortional bending stress for Example 7.1 with various diaphragm spacings

Diaphragm spacing: m	2	4	6	9	11	15	22
βL_D	0.34	0.68	1.0	1.6	1.9	2.7	3.8
f_{DW}: N/mm^2	2	8	18	45	45	45	45
f_{DB}: N/mm^2	0	0	1	5	10	34	34

To determine the longitudinal distortional stress the various coefficients are determined:

$$\frac{DD_{YT}}{BD_{YW}} = \frac{2 \times 3.8}{0.9 \times 0.14} = 60$$

$$\frac{D_{YB}}{D_{YT}} = \frac{3.8}{3.8} = 1$$

From Fig. 7.2(a), $k_3 = 0.12$

$$k_2 = \frac{24 D_{YT} k_3}{B^3} = \frac{24 \times 3.8 \times 0.12}{0.9^3} = 15$$

Also, $\beta L_D = (k_2 L_D^4 / EI)^{0.25} = (15 L_D^4 / (210\,000 \times 0.81))^{0.25}$, and solving this gives $\beta = 0.17$.

Using Equation (7.6) the longitudinal warping stress can be calculated for various diaphragm spacing. From Table 7.1 it can be seen that the longitudinal distortional warping stress increases with diaphragm spacing, but is constant beyond $\beta L_D = 1.6$. From Fig. 7.4 the diaphragm spacing near midspan is 6 m and $f_{DW} = 18$ N/mm^2. The peak longitudinal stress is the sum of the bending, distortion and warping stresses (Equation (7.8)):

$$f = f_b + f_{TW} + f_{DW} = 250 + 6 + 18 = 274 \, \text{N/mm}^2$$

The limiting stress for a 60 mm thick flange plate is $0.95 f_y = 308$ N/mm^2, again this is satisfactory.

For the transverse distortional bending effects, again the relative rigidity of the web and flanges is used to determine coefficient k_6 from Fig. 7.2(b) as 0.25, and k_5 as 0.113. Using Equation (7.9) the values of the transverse bending stress are tabulated for various diaphragm spacing. From Table 7.1 it can be seen that the transverse distortional bending increases with diaphragm spacing but is constant beyond $\beta L_D = 2.65$. For the 6 m diaphragm spacing used in this example the transverse stresses are small.

The transverse bending stresses will need to be carried around the welded joint between the flange and web; these stresses will be in addition to any local bending from the application of the load and any longitudinal shear stresses. For a railway structure of this type the use of a full-penetration butt weld is advisable to avoid fatigue issues (Fig. 7.5(a)). For highway structures the welds may be reduced to partial-penetration welds (Fig. 7.5(b)), but a double-sided weld should be used to avoid bending of the weld itself. In both details the primary weld is formed from the outside of the box to limit work inside the box.

The stresses in the bridge were also checked using a finite element analysis. The maximum bending stresses are similar to those estimated from the equations. The

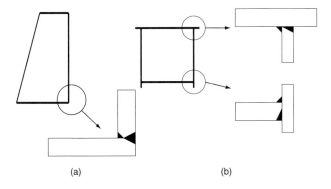

(a)   (b)

Figure 7.5 Weld details for box connections: (a) standard railway box; (b) composite highway box.

web shear stresses are higher near the support, but this is not related to the box behaviour. The stresses are significantly different to the average stresses assumed for design; for larger boxes with limited post-buckling capacity this elastic shear distribution may become critical and additional stiffeners may be required to limit buckling at serviceability limit states.

A review of the behaviour of composite box girders indicates that the stresses induced are a result of the relative flexibility of the steel plate elements of the box. The use of a composite top flange will increase the rigidity of this element. Reducing the bottom flange to the bare minimum size then leads to a triangular shaped structure. This form is stiffer and less susceptible to distortion than the conventional rectangular or trapezoidal forms. Where a conventional box shape is required the rigidity of the web can be significantly increased by the use of a folded plate. The folded plate gives a shear and torsional stiffness similar to that of a conventional web, but transverse rigidity is increased by a factor proportional to the square of the fold amplitude (a). The folded plate effectively provides a continuous but relatively weak ring diaphragm; this is a valid option to the practical diaphragm stiffness requirements at low βL_D ratios (Fig. 7.3).

The Maupre Bridge in France [18] is an example of a structure where the triangular box and folded web have been exploited to remove all intermediate diaphragms (Fig. 7.6). The behaviour of a folded plate web will be different from a conventional web; the increased distortional stiffness is accompanied by a reduction in longitudinal stiffness, such that the web can carry almost no longitudinal bending. Further consideration of folded web plates is set out in Chapter 11. A number of innovative composite box girder forms have been developed in France [18] particularly on the various sections of the high-speed railway [92]. The development of composite sections for high-speed railways serves as an interesting example of the strengths of this structural form.

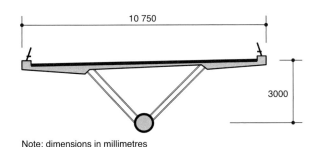

10 750

3000

Note: dimensions in millimetres

Figure 7.6 Details of the Maupre Bridge combining a triangular form with a folded web.

Example 7.2

The construction of high-speed railways across Europe has required the construction of many bridges and viaducts in various forms. In the UK the major structures have tended to be concrete box girders (Fig. 7.7(a)) [85] with only a few steel–concrete plate girders being used (Fig. 7.7(b)). This multiple girder structure has also been used in Europe [92], it has advantages in that the girders are directly beneath the rails. The use of a twin box-girder section (Fig. 7.7(c)) increases the stiffness of the section particularly for skew crossings. The inclined webs give improved aesthetics; however, the twin boxes are still relatively expensive to fabricate and erect.

Through the longer lengths of high-speed railway constructed in France and Spain, other forms of composite bridge have developed. The twin girder (Fig. 7.7(d)) increases the structural efficiency (see Chapter 5) and is simpler to build, particularly if launched. However, for the twin girder, the load transfer is less direct, involving slab bending; for asymmetric loading this involves some transverse rotation of the passing train. Deflection and rotational limits for high-speed lines are more onerous than conventional railways (Table 4.2) and often govern the sizing or layout of a bridge. To improve the rotational stiffness of the twin

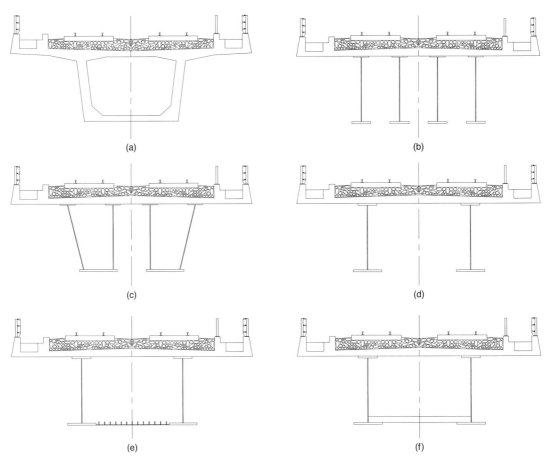

Figure 7.7 High-speed railway viaduct forms: (a) concrete box; (b) steel girders; (c) twin boxes; (d) twin girder; (e) composite box; (f) double composite box.

girder structure the addition of a lower steel plate between girders (Fig. 7.7(e)) has been used to form a box with its improved torsional inertia. On more recent structures the use of a lower concrete slab between the girders (Fig. 7.7(f)) has been used to form the box structure, this also improves the acoustic performance of the structure. The slab is not continuous longitudinally at midspan, but the panel lengths are sufficient to allow torsional shears to be transferred. This form gives the relatively simple fabrication and construction form of the twin girder bridge with the improved rotational and acoustic properties of the concrete box. For a given span the vertical deflection under load is broadly similar for all forms. The rotational stiffness of the box forms is significantly enhanced with rotations about three times less than the girder forms for the single-loaded-track scenario.

Noise from bridges

The construction of new railways and highways is often constrained by environmental issues such as noise. Noise can be mitigated by the choice of an appropriate bridge design. The noise associated with a road or railway tends to come from three sources: the vehicles or rolling stock, the wheel–rail or wheel–road interface and re-radiated structure-borne noise.

The noise from vehicle or rolling stock, motors, air-conditioning, and so on is largely controlled by the performance specification of the vehicles; the bridge form has no influence on this noise source. Wheel–rail or wheel–road interaction noise is related to the roughness of the road or railway; the noise generally increases with the age of the road or railway, for most situations this is the primary source of noise. Noise from points, switches or joints at the ends of bridges also contributes to this directly radiated noise. The radiated noise from wheel–rail interaction and vehicle or rolling stock can be controlled by the use of noise-absorbing barriers. For railways the barriers are more effective if placed adjacent to the track. For both rail and road bridges the use of large structural parapets should be used with care as the increased area may increase the re-radiated noise [85].

Structure-borne noise is generated from vibrations from the wheel–rail interface, which are transmitted through the structure and re-radiated. This is largely a dynamic phenomenon depending on the mass, stiffness, natural frequency and damping of the structure [76]. Massive or highly damped structures emit less noise. The exposed surface area of the structure also affects the amount of noise generated, and this should be minimised where possible. Figure 7.8 outlines the relative noise generated by a high-speed train on the various viaduct forms of Fig. 7.7. The box sections are less noisy than the girder forms. The concrete box and double composite steel–concrete box are the best at limiting re-radiated noise. For the steel composite forms the webs are the major sources of noise, the natural frequencies tend to be in the range generated by the train leading to some resonance. The use of thicker webs, additional stiffening or damping can be used to reduce this noise.

Analysis of the radiated noise is a relatively complex procedure. An initial estimate of the relative noisiness of various structural forms can be made by assessing the vibration frequency of the primary elements. To estimate the actual noise generated involves significantly more complex analysis. The initial problem is to estimate the input energy; this will require modelling track and wheel defects. Having defined the input criteria the behaviour of the structure must be modelled,

93

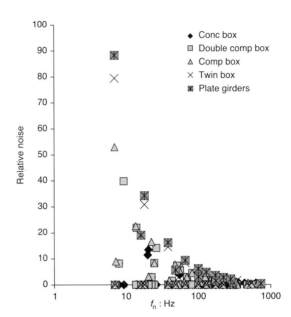

usually by finite element techniques. The modelling required to achieve frequency outputs to about 300 Hz requires a model more refined than is usual for stress analysis [86].

Shear connectors for composite boxes

For an open topped box (Fig. E1(b) of Appendix E), the behaviour of the steel–concrete interface is similar to that of a conventional girder and the shear flow is calculated from Equation (1.16) (but taking into account the additional torsional shears). The shear failure planes will be the same as those of Fig. 1.9. For a closed box (Fig. E1(a)) of a similar size, logic indicates that a similar number of connectors should be satisfactory; this is generally true, but with some complications. The top flange layout is not symmetrical about the web and so the connectors previously provided outside the web are added inside, increasing the connector density locally near the web. The majority of the effective connectors (approximately 90%) should be within the effective width of the girder, connectors outside the effective width will be required to carry local or transverse effects or the small longitudinal shear spread beyond the effective width. The nominal connector spacing requirements (Table 1.3) also apply in this area.

The load in any connector at a distance a from the web of a closed box is given by:

$$P = \frac{Q_1}{No}\left\{ k\left(1 - \frac{a}{b_{\mathrm{w}}}\right)^2 + 0.15 \right\} \tag{7.10}$$

where Q_1 is the longitudinal shear derived for one web, b_{w} is the width of the box from the web to the box centre line, No is the number of connectors within b_{w} (all are assumed to be of the same type), k is a coefficient taken from Fig. 7.9. The number of connectors within a nominal 200 mm strip adjacent to the web is No'.

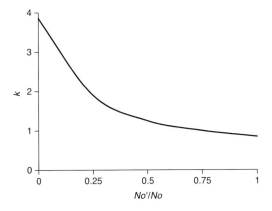

Where connectors are not of the same type the effective number of connectors may be estimated as follows:

$$No = No_1 + No_2 \frac{P_{u_2}}{P_{u_1}} \tag{7.11}$$

Composite plates

The top flange of a closed box girder will initially provide support for the wet concrete; it will also assist the embedded slab reinforcement to carry local wheel loads. The flange can be considered as providing a tie restraint for arching action on smaller spans (see Chapter 5), or as providing additional bending reinforcement for larger spans.

During construction of a box girder the top flange will be subject to a direct in-plane compression (or tension) induced in the flange by the bending of the box section. To prevent buckling of the flange it should satisfy the shape limitations outlined in Table 1.3 unless stiffening is added. The plate will also be subject to direct loading from the wet concrete; the deflection of the plate should be limited in this situation. For a 200 to 300 mm-thick deck slab the deflection should be limited to the span/300 (Equation (4.12)) or 10 mm, whichever is less. A plate thickness of approximately the span/60 will be required for unstiffened plates; where the span is larger then stiffening of the plate will be required.

For the design of the composite plate to resist local wheel loads, two design methods are used. The simplest method is to ignore the capacity of the plate and provide embedded reinforcement to carry the local wheel loads; shear connectors at a nominal spacing only are provided to ensure the plate and concrete remain together. This method is conservative but useful particularly on multiple box structures where only part of the slab soffit has a composite plate. The second method considers the steel plate as reinforcement; the required area is derived using Equation (1.6). The longitudinal shear is estimated from Equation (1.16) and the spacing of the shear connectors is derived from Equation (1.17) as for other steel–concrete composite sections. Where the slab is subject to moments that cause compression in the plate the connector spacing should be such that local buckling of the plate does not occur (Fig. 7.10(b)). In practice, failure of short spans is often by punching shear (Fig. 7.10(c)); again unless the connectors are at an increased spacing the slab will not act as effective reinforcement for this failure type.

95

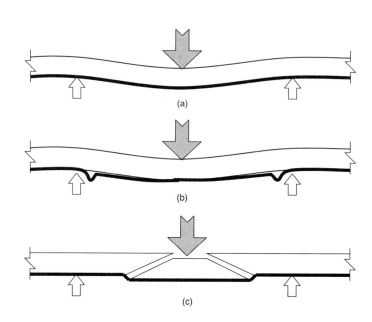

Figure 7.10 Failure mechanisms for composite plates [84]: (a) yielding of the plate in bending tensile zone; (b) buckling of the plate in bending compression zone; (c) plate ineffective for local shear failure.

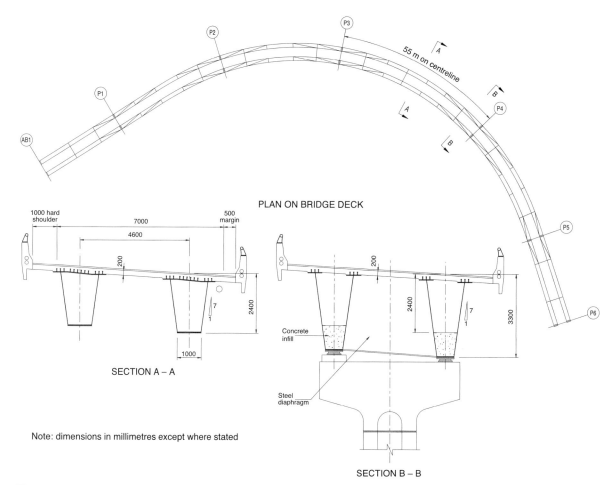

PLAN ON BRIDGE DECK

SECTION A – A

Note: dimensions in millimetres except where stated

SECTION B – B

Figure 7.11 Typical viaduct for curved interchange structure of Example 7.3.

Example 7.3

A series of viaducts consisting of twin trapezoidal box girders was proposed for new grade separated highway intersections in Kuala Lumpur. The road alignment was highly curved and pier positions restricted by existing roads and services; typically, 55 m spans were used (Fig. 7.11). At midspan a closed box, and nearer the supports an open box were proposed. At the piers the boxes were haunched and a lower concrete slab was placed within the box to form a double composite section (see Chapter 6).

Due to the curved alignment, closed-box behaviour was required during construction to carry the wet concrete loading, and a plan-bracing layout was used to achieve this. For the torsional analysis the bracing was assumed to act as a plate, the effective thickness of the plate being estimated from:

$$t_{eff} = \frac{A}{b} \cos\theta \sin^2\theta \qquad (7.12)$$

where A is the area of the brace, b is the distance between webs and θ is the brace angle.

Loads and analysis

For a 55 m span, the permanent load on the non-composite section is 2.6 MN, and the load on the composite section is 4.5 MN. The characteristic live load for a single lane giving maximum torsion is 2.1 MN. Analysis of the structure, using a grillage method, gives estimates of shear and torsion near the quarter span (where the section changes from an open to a closed box). The section properties and torsional properties are calculated as shown in Appendixes B and E.

For the box the effective shear on a web can be estimated by rearranging Equation (7.2):

$$V_T = \frac{T}{2B} \qquad (7.13)$$

For this example with $T = 1.6$ MNm and $B = 1.25$ m,

$$V_T = \frac{T}{2B} = \frac{1.60}{2 \times 1.3} = 0.67 \text{ MN}$$

$$Q_1 = (0.81 + (0.5 \times 1.02)) \times 0.24 = 0.32 \text{ MN/m or } 320 \text{ kN/m}$$

Using Equation (1.17) to estimate the number of connectors: $No = Q_1/0.55P_u = 320/(0.55 \times 108) = 5.4$ rounded up to 6, so pairs of connectors at 300 mm centres are sufficient. Using Equation (7.10) the distribution of force in the connector is as Fig. 7.12.

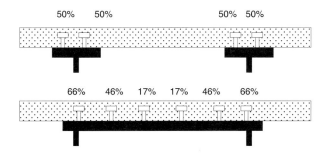

Figure 7.12 Distribution of force in connectors for the open and closed box of Example 7.3.

8

Trusses

...the simpler truss types ...with linked bracing between bays or bracing that is directly in line give the best forms ...

Introduction

The heyday of the truss was in the nineteenth century when many types were developed empirically (Fig. 8.1), all aimed at carrying the relatively heavy rolling load of the train. Nowadays, the relatively high fabrication costs of the truss members and in particular the numerous joints have meant that they also tend to be economic only for larger spans or heavier loads such as railways.

The truss tends to be of two basic forms: the through-truss where the railway passes between a pair of trusses, or as an underslung truss where the railway rides over the truss. Composite action in the truss may occur in a number of

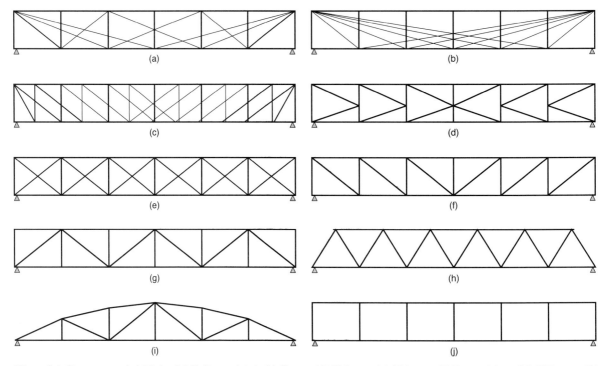

Figure 8.1 Truss types: (a) Fink; (b) Bollman; (c) double Pratt; (d) K-brace; (e) X-brace; (f) Pratt; (g) modified Warren; (h) Warren; (i) Bowstring; (j) Vierendeel.

Table 8.1 *Span–depth ratios for truss bridges*

Truss form	Typical span/depth ratio
Simply supported highway truss bridge	10–18
Continuous highway truss bridge	12–20
Simply supported railway truss bridge	7–15
Continuous railway truss bridge	10–18

ways. Usually the deck slab is formed of a concrete slab composite with steel transverse beams, the beams being supported by the trusses. This may be a filler joist form, or a beam and slab form. For the simply supported through-truss the slab is at the bottom chord level and so is in the tensile zone, it does not act efficiently as a composite section. The underslung truss is more efficient in this respect as the slab forms part of the top compression chord; it also avoids the additional transverse bending effects induced from the cross-girders in the through-truss form. Occasionally where a double-deck structure is used, both forms are combined. Trusses may also be used on continuous structures [87] or cable-stayed structures [88, 89].

Typical span-to-depth ratios for trusses are outlined in Table 8.1. Railway structures will tend to have lower span-to-depth ratios because of the higher live loads and the tight deflection and rotational limits of modern high-speed railways. The angle of the diagonals should generally be kept constant, to keep details standardised and to avoid a cluttered visual appearance. For almost any truss structure when viewed obliquely there will be an intersection of elements. Diagonals at 45–65° to the horizontal are generally the most efficient structurally; shallower angles tend to have higher relative loads and are less stiff. The Fink and Bollman trusses (Fig. 8.1(a) and (b)) that have relatively shallow diagonals tend not to be used nowadays (although they have produced some stunning structures [90, 110]). The bay length of a truss should be a multiple of the transverse beam spacing, an even number of bays being preferred to avoid crossed diagonals if a modified Warren or Pratt truss configuration is used.

Example 8.1

An analysis of the ten truss types shown in Fig. 8.1 has been carried out, assuming a 60 m span and a 10 m height. The structures were sized to carry a 1 MN rolling load. Both through and underslung forms were considered. The results of the analysis are shown in Fig. 8.2. A number of conclusions can be drawn, the most obvious being that all the braced forms of truss are significantly lighter than the Vierendeel type. It is also clear that the underslung system utilising a composite beam and slab system is more efficient than the through-truss for all truss types. In general the stiffer the structure for a given steel tonnage the more efficient the structure [91], the simpler truss types tending to be stiffer. Those trusses with linked bracing between bays (Fig. 8.1(h)) or bracing that is directly in line (Fig. 8.1(i)) give the best forms.

The truss forms were analysed, as is traditional, as pin-jointed frames. This is permitted by codes [2] as the moments at joints induced by deflection are secondary and not required for the equilibrium of external loads. Most well-framed structures have sufficient deformation capacity at joints for this assumption to be valid. Where node eccentricity is present (Fig. 9.9) then the moments should be considered;

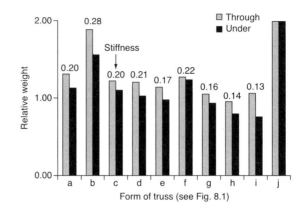

*Figure 8.2 Relative weight
and stiffness of various truss
types shown in Fig. 8.1.*

where truss members are stocky the moments may also be more significant, and of course for untriangulated forms such as the Vierendeel (Fig. 8.1(j)) the moments will predominate the design.

A number of simple criteria for the efficient design of truss bridges have been developed (based on [91]) and in order of relative importance are:

- Each bay of the truss should be braced (not Vierendeel).
- The truss should support the load directly (under-slung rather than through).
- The truss layout should exploit the properties of the materials used (composite with concrete in compression and steel in tension).
- The size of the bracing should reflect the loads carried (using smaller diagonals at midspan where shears are low).
- The bracing between bays should be linked (Warren truss rather than Pratt type).
- The bracing should be linked in a continuous line between bays (a bowstring form).
- The bracing should be at 45 to 65° to the horizontal (Warren truss rather than Bollman type).
- When additional bracing is added it should follow the above criteria (i.e. the bracing within the bowstring form).

From the criteria a steel–concrete composite, under-slung fish-belly or Warren form, emerges as the most efficient. Both forms have recently been used to support high-speed railways in Europe [92, 93]. Figure 8.3 outlines these two structures.

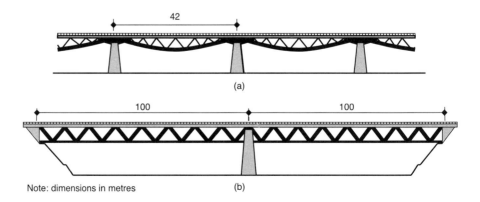

*Figure 8.3 Composite
under-slung truss:
(a) fish-belly at L'Arc;
(b) Warren truss at
Hammerbruke.*

Note: dimensions in metres

Member type

The type of member used in the truss will depend upon the span, loading and the force it is carrying (tension or compression). For smaller spans and lighter loads, rolled beam, channel or tubular sections may be appropriate, and for most bridge trusses a fabricated section will be used. The form of the fabricated section will depend upon the location of the member. For compression chords the effective length for buckling out of the plane of the truss is likely to be longer than in-plane, the use of an I section on its side is more appropriate to maximise the transverse radius of gyration. The vertical I section is more appropriate for the lower tensile flange (on a simply supported truss) that also carries local bending from the transverse beams. Open I, U or C sections are preferred for their easier fabrication and because they allow simpler splice plate connections. Box sections are efficient for larger trusses where internal access can be gained. The use of a concrete-filled section, creating a steel–concrete composite section, can be used to increase the capacity of the basic steel section.

Steel sections under axial load

In Chapter 1 the basic limitations of steel sections in buckling was outlined, the limiting axial compression load being determined from Equation (1.7) provided that local buckling is prevented by the limits to plate geometry (as Table 1.3):

$$N_D = 0.95 A_a f_{ac} \tag{1.7}$$

The limiting compressive stress depends primarily on the member's slenderness parameter λ; for axial loads this is dependent on the effective length (L_e) of the member and its radius of gyration (r):

$$\lambda = \frac{L_e}{r} \tag{8.1}$$

For design there is a variation in capacity for different section types (Fig. 8.4); small, hot-rolled, hollow sections having larger capacities than large, fabricated, welded sections with thick plates (greater than 40 mm). Rolled beam and column sections or fabricated sections using thinner plates have intermediate relative capacities.

Typical effective lengths are given in Table 1.4, for truss members. Generally the out-of-plane buckling of a truss member is the more critical, hence it is preferable to use sections with larger transverse inertia.

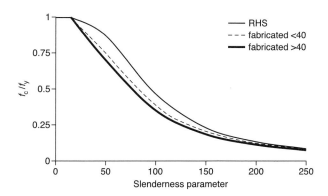

Figure 8.4 Limiting compressive stress for grade 355 steel for members subject to axial loads: (a) hot-rolled hollow sections; (b) fabricated sections with plates less than 40 mm thick; (c) fabricated sections with plates more than 40 mm thick.

Joints in steelwork

A steel truss bridge is generally made up of a series of chord and diagonal components that are joined together. The joints may form the weakest element in the truss, at nodes the stresses may be complex with bending and axial loads from various directions being combined and resolved. Generally jointing should be carried out away from the truss nodes, particularly if there is a change in section type (box to I or vertical to transverse I). Two forms of joints in steel structures are common: fully welded joints, or bolted joints. The choice of joint type will be dependent on a number of factors, particularly the expertise of the contractor and fabricator building the structure, the form of the structure, and the number and size of joints.

Welded joints

Welded joints are suitable for in-line flange- or web connections where the full capacity of the section needs to be maintained. Welded joints are also suitable for structures where a protected environment can be created for the fabrication of a number of joints, that is viaduct structures (Chapter 5) or launched bridges. Full-strength joints will require the use of full penetration butt welds; where there is a change in plate thickness on one side of the joint the plate should be tapered at a minimum of 1:4 to limit stress concentrations and maintain a reasonable fatigue strength.

Bolted joints

For bridges, all bolted connections between main structural sections should use high-strength, friction-grip (HSFG) bolts. These bolts are preloaded and have a low initial slip giving a rigid joint with good fatigue properties. They are best used in shear using lapped joints or joints with cover plates (Fig. 8.5(a) and (b)). The use of end plates where bolts are in direct tension should be avoided as the connection is more flexible, is subject to prying forces and is generally more difficult to fabricate (Fig. 8.5(c)).

The strength of a joint formed from HSFG bolts will be determined by the lower of: (i) a failure in the elements being connected (usually across the bolt holes as a

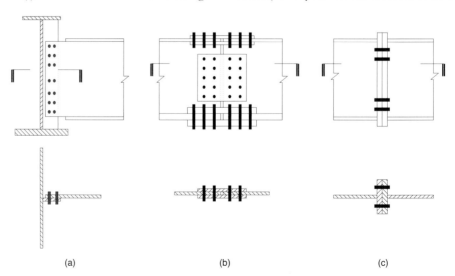

Figure 8.5 Bolted joint types: (a) lap joint in single shear; (b) cover plates in double shear; (c) plate joint in tension.

(a)　　　　　　　　　　(b)　　　　　　　　　　(c)

Table 8.2 HSFG bolt properties

Bolt	Diameter d_b: mm	Tensile area A_{be}: mm^2	Tensile stress: N/mm^2	Proof load F_o: kN	Hole diameter: mm
M20 general grade	20	245	587	144	22
M20 high grade			776	190	22
M24 general grade	24	353	587	180	27
M24 high grade			776	273	27
M30 general grade	30	561	512	287	33
M30 high grade			776	435	33
M36 general grade	36	817	512	418	39
M36 high grade			776	633	39

tension failure (P_T)), (ii) a failure in the cover plates (again usually across bolt holes (P_T) or by failure in bearing (P_B)), or (iii) of a failure of the bolts in shear (P_v). At the ultimate limit state these capacities may be determined from:

$$P_T = 0.95 f_y A_{ae} \qquad (8.2)$$

where A_{ae} is the area of the element being connected or of the cover plates, less the area of any bolt holes, and f_y is the yield strength of the element or cover plate.

$$P_B = k_2 k_3 k_4 0.9 f_y d_b t \qquad (8.3)$$

where k_2 is a coefficient depending on edge distances of the bolts, this is 2.5 for distances of $3d_b$ or more and 1.7 for the minimum $1.5d_b$ edge distance, d_b is the bolt diameter, k_3 is 1.2 for an enclosed plate or 0.95 for a cover plate and k_4 is 1.5.

$$P_v = N_{o2} 0.64 f_{yb} A_b \qquad (8.4)$$

where N_{o2} is the number of shear planes (Fig. 8.5), f_{yb} and A_b are the yield stress and area of the bolt (Table 8.2).

For HSFG bolts the frictional capacity (P_F) of the connection will often govern the design, at the serviceability limit state the bolts should not slip:

$$P_F = k_1 N_{o1} 0.83 F_v \mu \qquad (8.5)$$

where k_1 is 1.0 for normal size holes and 0.85 for oversize holes, N_{o1} is the number of friction interfaces (usually the same as the number of shear planes), μ is the coefficient of friction that may vary from 0.23 for painted interfaces to 0.45 for steel to steel, $F_v = 0.85 F_O$, where F_O is the bolt proof load (see Table 8.2).

Example 8.2

The next example in this chapter looks at a steel through-truss bridge constructed to carry a high-speed railway into London. The structure spans 75 m and carries two tracks over the East Coast Main Line (ECML) to a tunnel portal. The bridge is enclosed with an architectural cladding and has a tubular shape. The track through the bridge is on an isolated track slab to limit noise and vibration.

Enclosure

Steel requires both water and oxygen to corrode (Chapter 3), and enclosure of the steel structure with a waterproof enveloping structure is an alternative to painting or the use of weathering steel. Enclosure may also be used as a supplementary protection layer to give enhanced durability for important structures [74]. To be successful the enclosure must keep out water and vermin, but have adequate ventilation to prevent high moisture content in the enclosed air space. The enclosure must also not impede regular inspection of the structural elements and must be simple to maintain itself and be replaceable relatively easily (the life of an enclosure system is likely to be 30 to 50 years). Enclosure is a relatively expensive way to protect the steelwork from corrosion, but where an enclosure is proposed for other reasons, for example architectural, fire or noise protection, then the additional cost may be offset.

For the structure of Example 8.2 the cladding serves a number of functions. The structure is located at a tunnel portal over the ECML, a robust structure with enhanced fire protection was deemed necessary. Thick steel sections are used as the primary protection (thick steel sections are less likely to buckle or fail in a fire than thin sections), a fire-resistant cladding is also provided to give added protection. The central London location also imposes environmental issues including the need to limit railway noise; the internal cladding also assists this. The complete enclosure reduces direct rail and locomotive noise as well as the reradiated structure noise (see Chapter 7). The cladding support structure is isolated from the main truss structure by steel–elastomeric units to limit the transfer of vibration to the cladding, limiting the reradiated or 'rumble' noise. The external panels provide a curved architectural cladding and waterproof skin.

The structural form is a pair of Warren girders with chords and diagonals formed from fabricated girder sections. The top chord and diagonals have the beams oriented transversely to maximise the resistance to lateral buckling (Fig. 8.6(a)). The lower chord is orientated vertically to resist the local bending from train loads and the significant local reactions induced during the launched construction of the bridge (Fig. 8.6(b)). Cross-beams span transversely between the lower chords and support a concrete structural slab. Plan bracing spans transversely between the top chords.

Loading and analysis

The total weight of the bridge (G), cladding and fixings is 25 MN. Total live loading (Q) is similar. The bridge analysis is carried out using a three-dimensional truss model with continuous joints such that moments, shears and axial loads are

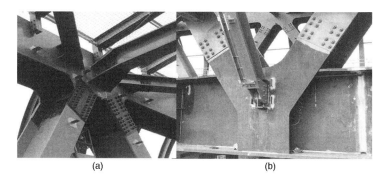

(a)   (b)

Figure 8.6 Chord to diagonal intersection for Example 8.2: (a) upper node; (b) lower node.

obtained; however, for the primary member design and splice design the moments are small and only primary, axial load will be considered.

The load in the chords of a truss may be estimated by assuming that the compression and tension in the chords resists the bending moment calculated by assuming that the truss is a conventional girder:

$$N_C = N_T = \frac{M}{D} \tag{8.6}$$

Similarly the forces in the diagonals can be calculated by resolving the calculated shear force through to the members' inclination:

$$F = \frac{V}{\sin \theta} \tag{8.7}$$

For this example at the ultimate limit state, $M = 600$ MN m and $V = 32$ MN. The depth between the centre of the chords is 7.58 m and the angle of the diagonals to the horizontal is 58°. Using Equation (8.6) the compression in each of the two top chords is:

$$N_C = \frac{M}{D} = \frac{600}{7.58 \times 2} = 40 \text{ MN}$$

From the analysis model the compression is 37.2 MN and there are some associated moments.

The effective length of the top chord from Table 1.4 is 0.85 times the length between truss nodes for both in-plane and out-of-plane buckling, $L_e = 0.85 \times 9.4 = 7.9$ m. Section properties for the chord are outlined in Appendix C, the in-plane radius of gyration being the smaller with r_y being 0.29 m. Using Equation (8.1):

$$\lambda = \frac{L_e}{r} = \frac{7.9}{0.29} = 27$$

From Fig. 8.6 using the curve for thick fabricated sections, $f_{ac} = 0.87 f_y$. For sections with 65 mm plate, $f_y = 315$ N/mm^2 (Fig. 1.5). Using Equation (1.7) with $A_{ae} = 0.225$ m^2, the capacity of the section is:

$$N_D = 0.95 A_a f_{ac} = 0.95 \times 0.87 \times 315 \times 0.225 = 58.7 \text{ MN}$$

This is larger than the applied 40 MN.

The force (compression) in the diagonals adjacent to the support at the ultimate limit state, using Equation (8.7) is:

$$F = \frac{V}{\sin \theta} = \frac{32}{2 \times \sin 58°} = 19 \text{ MN}$$

At the serviceability limit this is reduced to 13.7 MN.

At both the upper and lower nodes the joint is a fully welded connection with radiused plates to reduce stress concentrations (Fig. 8.6(a) and (b)). A bolted connection, formed using splice plates with HSFG bolts loaded in shear, is used just away from the node.

The number of bolts in the section may be approximated initially by dividing the force by the bolt capacity:

$$No = \frac{N_F}{P} \tag{8.8}$$

where P is the appropriate bolt capacity in shear, bearing, tension or friction.

Assuming M36 HSFG bolts will be used with $A_b = 817\,\text{mm}^2$ and $f_{yb} = 512\,\text{N/mm}^2$ (Table 8.2), then from Equation (8.4) for bolts in double shear:

$$P_V = N_{o2}0.64f_{yb}A_b = 2 \times 0.64 \times 512 \times 817 \times 10^{-3} = 535\,\text{kN}$$

The maximum number of bolts required at the ultimate limit state is:

$$No = \frac{N_F}{P_V} = \frac{19}{0.535} = 36$$

The diagonal section consists of unpainted steel with a steel-to-steel faying surface, $\mu = 0.45$, $F_o = 418\,\text{kN}$ and $F_v = 355\,\text{kN}$. Using Equation (8.5):

$$P_F = k_1 N_{o1}0.83F_V\mu = 1 \times 2 \times 0.83 \times 355 \times 0.45 = 265\,\text{kN}$$

$$No = \frac{N_F}{P_V} = \frac{13.7}{0.265} = 52$$

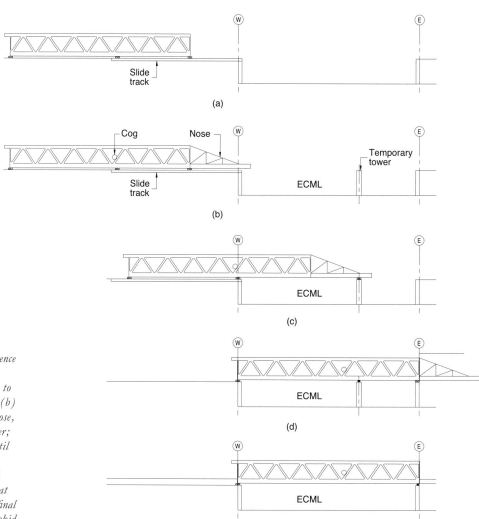

Figure 8.7 Launch sequence for truss bridge: (a) construct bridge adjacent to railway on launch track; (b) place launch skids and nose, construct temporary tower; (c) launch the bridge until the nose is on the tower, remove central skid; (d) launch bridge until it is at final location; (e) place final bearings, remove launch skid and nose.

The number of bolts is governed by the serviceability requirements in this case, to limit slip. This is generally true for most HSFG joints on bridges.

The cover plate sizes are smaller than the main sections and so a tension check is made through the plates. The area of the plates is $107\,000\,\mathrm{mm}^2$, there are fifteen 39 mm-diameter holes across the width reducing the plate area (A_{ae}) to $77\,500\,\mathrm{mm}^2$. Using Equation (8.2):

$$P_T = 0.95 f_y A_{ae} = 0.95 \times 345 \times 0.0775 = 25\,\mathrm{MN}$$

The outer plate is a cover plate so $k_3 = 1.2$, and the holes are at the minimum edge distance so $k_2 = 0.95$. Using Equation (8.3):

$$P_B = k_2 k_3 k_4 0.9 f_y d_b t$$

$$= 0.95 \times 1.2 \times 1.5 \times 0.9 \times 345 \times 36 \times 2 \times 25 \times 10^{-3}$$

$$= 955\,\mathrm{kN}$$

For 61 bolts the capacity is 58 MN. The capacity of the splice in bearing and tension is above the applied force and is satisfactory.

The bridge is constructed adjacent to the railway and launched across it (Fig. 8.7). The behaviour of the bridge is significantly altered by this launch, tensions are in the upper (previously the compression) chord and compressions are in the lower (previously the tension) chord. More significantly, large local patch loading of the lower chord occurs.

Local loading of webs

When large localised loading occurs on a steel section that is unstiffened, then a local crippling of the web may occur. If the local bearing stress at the foot of the web is greater than a limiting stress (f_{ab}) then either stiffeners should be added or a more detailed check of the web crippling should be carried out.

$$f_{ab} = \frac{3 f_y t_w}{(a_1 d_w)^{0.5}} \tag{8.9}$$

For thin plates the provision of stiffeners is generally recommended. For thicker plates the limits of Equation (8.9) are conservative. For a web under a local patch load the lower of the local yield (R_{ay}) or the local buckling (R_{ab}) capacity should be used:

$$R_{ay} = \left(2 t_f (B t_w)^{0.5} + t_w a\right) 0.95 f_y k \tag{8.10}$$

$$R_{ac} = \left(0.45 t_w^2 \left(E_a \frac{f_y t_f}{t_w}\right)^{0.5}\right)\left(1 + \frac{3a}{D_w}\left(\frac{t_w}{t_f}\right)^{1.5}\right) k \tag{8.11}$$

where k is a reduction factor dependent on the stress in the flange:

$$k = \left(1 - \left(\frac{0.95 f_f}{f_y}\right)^2\right)^{0.5} \tag{8.12}$$

For Example 8.2 the launch reaction on one truss chord is approximately 10 MN at the ultimate limit state. The stress in the flange due to cantilevering of the truss and nose (during stages b to d of Fig. 8.7) is $140\,\mathrm{N/mm}^2$. The length of

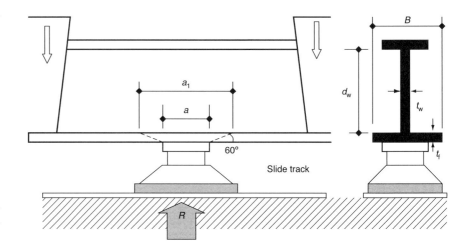

Figure 8.8 Patch loading on the web of Example 8.2, due to launching.

the local load a is 1000 mm and the force will spread through the flange at 60° (Fig. 8.8):

$$a_1 = a + 2\sqrt{3}t_f = 1000 + (2 \times 1.73 \times 65) = 1225\,\text{mm}$$

Using Equation (8.9):

$$f_{ab} = \frac{3f_y t_w}{(a_1 d_w)^{0.5}} = \frac{3 \times 325 \times 60}{(1225 \times 1635)^{0.5}} = 41\,\text{N/mm}^2$$

The bearing stress is

$$\frac{10 \times 10^6}{1225 \times 60} = 136\,\text{N/mm}^2$$

The bearing stress is significantly greater than the limiting stress f_{ab} and so the web crippling will be checked. Using Equation (8.12) the reduction factor k is:

$$k = \left(1 - \left(\frac{0.95 f_f}{f_y}\right)^2\right)^{0.5} = \left(1 - \left(\frac{0.95 \times 140}{325}\right)^2\right)^{0.5} = 0.91$$

Checking the web yielding using Equation (8.10) with $B = 1300$ mm:

$$R_{ay} = \left(2t_f (Bt_w)^{0.5} + t_w a\right) 0.95 f_y k$$
$$= \left(2 \times 65(1300 \times 60)^{0.5} + 60 \times 1000\right) 0.95 \times 325 \times 0.91 \times 10^{-6}$$
$$= 27\,\text{MN}$$

Checking buckling using Equation (8.1) using $t_w = t_f$:

$$R_{ac} = \left(0.45 t_w^2 \left(E_a \frac{f_y t_f}{t_w}\right)^{0.5}\right)\left(1 + \frac{3a}{D_w}\left(\frac{t_w}{t_f}\right)^{1.5}\right)k$$

$$R_{ac} = \left(0.45 \times 60^2 \times (210\,000 \times 325)^{0.5}\right)$$
$$\times \left(1 + \frac{3 \times 1000}{1635}\right) \times 0.91 \times 10^{-6} = 34\,\text{MN}$$

Both buckling and yield capacities are significantly greater than the applied loads; the local launching stresses are acceptable without additional stiffeners.

Continuous trusses

The basic trusses considered have been simply supported structures. Trusses can be used for continuous structures, often of large span. In the USA many large variable-depth, continuous trusses have been built using steel, the depth varying with the moment such that an approximately constant chord size can be utilised. Rearranging Equation (8.6):

$$D = \frac{M}{N_a} \tag{8.6b}$$

In Europe a number of continuous trusses have been used, particularly for railway structures [17, 87, 92, 93], the tendency has been for composite or double composite structures (Fig. 8.9) giving increased efficiency:

$$D = \frac{M}{N_{a-c}} \tag{8.6c}$$

Under-slung continuous trusses tend to be deep structures allowing the use of double-deck structures with two levels of highway or a highway at one level and a railway at another. For longer spans the trusses may be cable-stayed [88, 89]. The Øresund project in Sweden [87] utilised a series of 144 m span double-deck trusses with a 490 m cable-stayed section, the bridge carried a concrete deck supporting the road at high level with a railway on the lower chord. The Øresund Bridge also utilises high-strength steel.

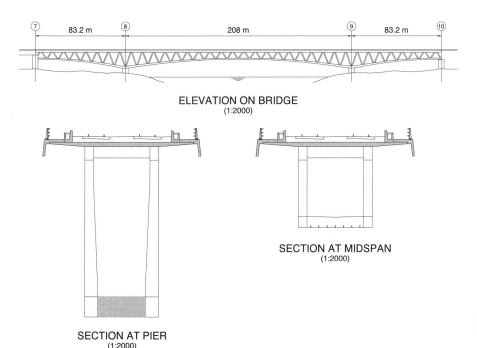

ELEVATION ON BRIDGE
(1:2000)

SECTION AT MIDSPAN
(1:2000)

SECTION AT PIER
(1:2000)

Figure 8.9 Variable depth continuous truss of the doubly composite Nantenbach Bridge.

High-strength steel

Typically in Europe steel with a nominal $355\,\text{N/mm}^2$ yield strength is assumed for the design of rolled sections and steel plate; most of the examples used to date have used this grade. Higher strength steel plate is available and a $460\,\text{N/mm}^2$ grade is now becoming more common (see Example 10.2). In Japan steel plate strengths to $700\,\text{N/mm}^2$ have been used. A number of issues need to be considered when using high-strength steel.

The first issue is weldability and fatigue. High-strength steel requires careful consideration of weld details. The steel is more difficult to weld and the fabrication shop must be set up to accommodate it. Standard production tests must be repeated for the higher grade, and this is unlikely to be economic for small structures. The higher stresses in the section will require additional testing [56]. The higher stresses used will also lead to fatigue being more critical, as the limiting fatigue stress (f_A) for various details is not increased with steel grade. The bridge should be designed with this in mind and details such as doubler plates with a low fatigue classification avoided. Second, the reduced steel area obtained for a given load will lead to a reduced stiffness, the elastic modulus for steel (E_a) is independent of strength. The increased stresses may also cause more cracking at continuous supports. Finally for steel sections subject to compressive loads the critical load at which buckling occurs will be affected. The slenderness parameter λ is modified. The slenderness parameter becomes

$$\lambda\left(\frac{f_y}{355}\right)^{0.5} \tag{8.13}$$

This means that for high-strength steel, buckling becomes more critical.

9

Arches

...sculpting bridges...

Introduction

The arch form has been around for millennia; many masonry arches (and some using concrete) survive from the Roman period 2000 years ago [97]. The arch carries load using compression, it is ideal for materials like concrete with limited tensile strength. Concrete is heavy and for large spans the higher strength-to-weight ratio of steel can significantly reduce arch thrusts. The problem with steel arches is that the steel is prone to buckle under compression and requires extensive stiffening and bracing. A composite structure if carefully considered can combine the best features of concrete and steel. Figure 9.1 outlines the typical arch forms and the terminology used. The use of crown, springing, rise and spandrel is derived from the terms used for masonry structures. Table 9.1 outlines typical geometric ratios used to size this form of structure.

The thrust of an arch (N_A) may be estimated approximately by calculating the equivalent bending moment that would occur on a beam of the same span (provided the arch shape and bending moment shapes are similar), then assuming that the lever arm resisting the moment is simply the arch rise (r). For parabolic arches under uniform loads this method gives good agreement with more complex

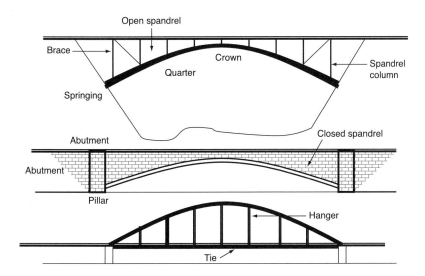

Figure 9.1 Arch forms and notation.

Table 9.1 Typical geometric ratios

Arch type	Span to rise: L/H
High	1.0
Semi-circular	2.0
Typical range	2.5 to 7
Flat	below 10

analytical methods:

$$N_A = \frac{M}{r} \tag{9.1}$$

Example 9.1

The first example in this chapter is used to explore the behaviour of a composite arch. The example considered is the Runnymede Bridge [96] over the river Thames, about 5 km downstream of Brunel's famous shallow, red brick arch, and a similar distance upstream of Hampton Bridge. The aesthetics of the structure were outlined in the 1930s by Edwin Lutyens to rival the other arches, including his 'bridge better than any others' [98] at Hampton. The structure has the appearance of a white stone arch with red brick spandrels. The external appearance is a wonderful example of this form [100]. The structure is formed from a thin, 225 mm depth of concrete slab spanning 55 m. It is stiffened by and acts compositely with steel truss ribs that support the upper deck slab, which carries traffic loads. The structure is now incorporated into the M25 carrying the northbound carriageway and the A30 with another new structure adjacent to it [99].

Loading and analysis

For the Runnymede Bridge the key dimensions are shown in Fig. 9.2. The total dead plus surfacing characteristic load is about 70 kN/m per lane. Assuming a 30 kN/m per lane live load we have a 100 kN/m per lane total load, say, about 145 kN/m at the ultimate limit state. For a span of 55 m the applied load and resultant forces are:

$G = 55 \times 145$ which gives 8.0 MN

The moment $M = 8.0 \times 55/8 = 55$ MN m at the crown. The rise of the arch is 5.5 m so the midspan thrust (using Equation (9.1)) is:

$$N_A = \frac{M}{r} = \frac{55}{5.5} = 10.0 \text{ MN}$$

The slab thickness is 225 mm, the cube strength of the concrete is 40 N/mm^2 so the arch capacity (using Equation (1.4)) is:

$$N_D = 0.4f_{cu}bh = 0.4 \times 40 \times 3.5 \times 0.225 = 12.6 \text{ MN}$$

Therefore, since $N_D > N_A$, the slab can carry the entire structure. However, traffic loading is seldom uniform in practice and the bridge will have to carry large, abnormal vehicles. The moment diagram of the abnormal vehicles is not the same shape as the arch (particularly for loads placed near the quarter span) and the resultant thrust occurs outside the arch. Large moments are induced in the slab together with large deflections, and the slab will buckle unless stiffened (see Fig. 9.14).

The Runnymede Bridge slab is stiffened by steel ribs at approximately 1.6 m centres. The ribs, that vary from 800 mm-deep plate girders at midspan to 5 m-deep trusses near the springing, carry the moments from asymmetric loads. The steel elements themselves are encased in concrete to form a double composite action (Chapter 6), which helps to carry secondary, local moments more efficiently. The moment is carried by the ribs and resisted as a couple by the top and bottom elements.

Another major problem with concrete arches is that they require significant falsework support during construction; this centring will fill the entire span and is often a significant structure in its own right. At Runnymede, the problem was overcome by using the steel stiffening ribs as the support during construction. The steel rib trusses were built in cantilever from each side of the river. Once the steelwork was closed, formwork for the arch was hung from the steel and the slab concreted in bays, leapfrogging from each springing to crown. This method of construction (Fig. 9.3) complicates the stress analysis of the structure as some of the dead load is now carried by the steelwork rather than all of it being carried by the concrete slab as previously assumed.

The steel–concrete composite arch form using the steel to carry initial construction loads has been developed dramatically in China in recent years (Fig. 9.4), and spans of 425 m have been achieved using the technique. To reach these spans the structural sizes are larger [73, 101, 106]. The diameter-to-thickness ratios (D/t) of the tubular sections for these bridges are sometimes greater than current UK limits [102].

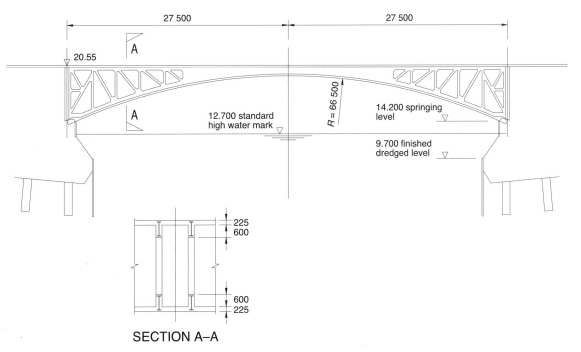

Figure 9.2 Runnymede Bridge details.

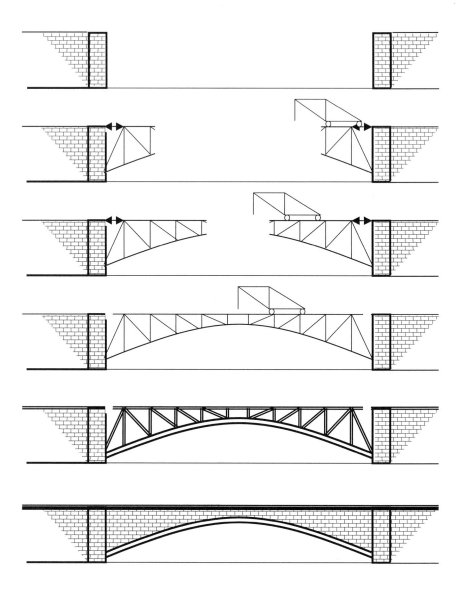

Figure 9.3 Runnymede Bridge construction sequence.

Composite compression members

So far in this and previous chapters we have been considering composite structures primarily in bending. The arch carries load primarily as direct compression. The maximum axial, or squash load carried by the steel–concrete composite section can be considered as the sum of its concrete, reinforcement and steel components, provided the steel element or connectors comply with the minimum thickness given in Table 1.4:

$$N_{ul} = 0.45 f_{cu} A_c + 0.95 f_y A_a + 0.87 f_{ys} A_s \tag{9.2a}$$

$$N_{ul} = 0.56 f_{ck} A_c + 0.95 f_y A_a + 0.87 f_{ys} A_s \tag{9.2b}$$

The relative contribution of the steel and concrete elements is measured by the contribution factor α. This is an important parameter for determining the likely behaviour of the composite section. When based on the concrete contribution, if

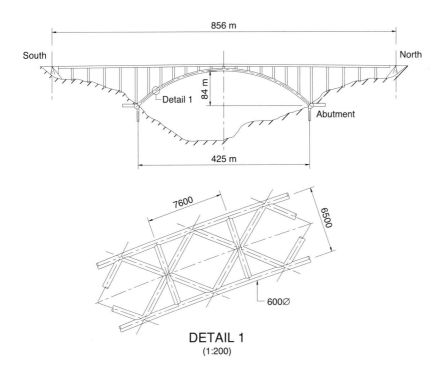

α_c is less than 0.1 then the section has minimal composite action and the section is designed as a steel section; as the ratio rises the concrete component of the structure becomes more important. Beyond 0.8 the concrete will dominate the section, with the steel elements acting in a similar way to reinforcement. The contribution factor may also be stated in terms of the steel contribution (α_a) [105]:

$$\alpha_c = \frac{0.45 f_{cu} A_c}{N_{ul}} \tag{9.3a}$$

$$\alpha_a = \frac{0.95 f_y A_a}{N_{ul}} \tag{9.3b}$$

When L_e/D_o is less than 12, the section can be considered as a short column and the ultimate capacity of the section taken as:

$$N_D = 0.85 k_1 N_{ul} \tag{9.4}$$

Where L_e/D_o is less than 4.5 the coefficient k_1 may be taken to be 1.0. For L_e/D_o ratios above 4.5, k_1 is derived from Fig. 9.5 and is dependent on the slenderness parameter λ':

$$\lambda' = \left(\frac{N}{N_{cr}} \right)^{0.5} \tag{9.5}$$

$$N_{cr} = \frac{(EI)_{a-c}}{L_e^2} \pi^2 \tag{9.6}$$

where $(EI)_{a-c}$ is the stiffness of the composite section. For slender columns with L_e/D_o greater than 12 an additional destabilising moment (M_e) should also be considered acting on the section. An L_e/D_o ratio of 65 is the approximate limit

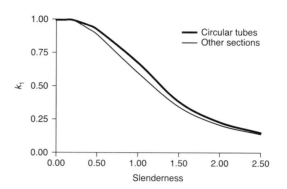

for this method:

$$M_e = 0.03D_o N \tag{9.7}$$

Considering a simple steel tube filled with concrete (Fig. 9.6), the maximum axial compression load is calculated when there is no moment on the section from Equation (9.4), there is a uniform stress distribution across the section. Similarly the tensile capacity of the section is calculated assuming a uniform stress distribution and ignoring any contribution from the concrete. With no axial load the bending resistance is calculated assuming a plastic stress distribution, with the compression and tensile forces being equal. If a small axial load is applied a small increase in moment capacity occurs; as more axial load is applied the moment will reduce, until the axial limit is reached with no moment. The values calculated are plotted to show the generalised form of the axial bending interaction curve (Fig. 9.7(a)). The capacity of the steel-only section is also shown on Fig. 9.7(a), from the curves it can be seen that the composite section has a greater axial and moment capacity compared to the steel-only section.

Figure 9.6 Stresses in a composite steel–concrete filled tube: (a) pure tension; (b) moment only; (c) combined axial and bending; (d) pure compression.

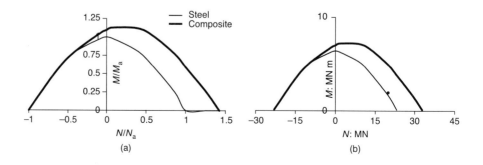

Figure 9.7 M–N interaction curves: (a) generalised comparing steel and composite sections; (b) curve for 900 mm tube of Example 9.2.

Example 9.2

The second example in this chapter is of an arch frame formed from a concrete-filled steel tube spanning the river adjacent to the previous structure, it is shown in Fig. 9.8. The structure was conceived to continue the aesthetic qualities of the other crossings. It uses a similar arch profile, but with an open structure to allow light to percolate through the total 100 m width of bridge at this location. The structure consists of a monocoque steel–concrete composite deck cantilevering from a central tubular arch frame. The frame consists of three 900 mm-diameter main tubes, with 600 mm-diameter bracing, the two lower tubes being concrete filled.

Loading and analysis

The total dead load of the new bridge is approximately 18 MN, with a maximum characteristic live load of 13 MN. Analysis of the structure uses a three-dimensional frame model of the combined deck and arch frame. The results of the analysis indicate that near the quarter span the lower composite tubes should be designed for a compression of 20 MN and a coexistent moment of 1.8 MN m, this moment including the additional moment induced due to the curvature of the section between nodes [103]. For the 900 mm-diameter tube filled with grade 45/50 concrete, using Equation (9.2b), and assuming that no additional reinforcement will be used:

$$N_{ul} = 0.56 f_{ck} A_c + 0.95 f_y A_a$$

$$= 0.56 \times 45 \times 0.555 + 0.95 \times 355 \times 0.078$$

$$= 14.0 + 26.0 = 40.0 \, \text{MN}$$

The steel contribution factor (Equation (9.3b)) is:

$$\alpha_a = \frac{0.95 f_y A_a}{N_{ul}} = \frac{26}{40} = 0.65$$

indicating that the steel contributes about two thirds of the section's strength.

Figure 9.8 Composite tubular arch at Runnymede.

The arch tubes are similar to the chords of a truss and the effective length will be approximately 85% of the distance between the bracing tubes. In this example the distance between nodes is 5.5 m. The section properties of the steel and composite section are given in Appendix C, $L_e = 0.85 \times 5.5 = 4.7\,\text{m}$, $I_{a-c} = 0.01\,\text{m}^4$, and using Equation (9.6) the critical load is:

$$N_{cr} = \frac{(EI)_{a-c}}{L_e^2}\pi^2$$

$$= \frac{210\,000 \times 0.01 \times \pi^2}{4.7^2} = 919\,\text{MN}$$

$$\lambda' = \left(\frac{N}{N_{cr}}\right)^{0.5} = \left(\frac{20}{919}\right)^{0.5} \quad \text{which gives } 0.2$$

from Fig. 9.5, $k_1 = 1$. The ratio $L_e/D_o = 5.2$, which is less than 12 and so no additional moment need be added. Using Equation (9.4) the axial compression capacity of the composite section is:

$$N_D = 0.85 k_1 N_{ul} = 0.85 \times 1.0 \times 40 = 34\,\text{MN}$$

The tensile capacity is 25.0 MN and the calculated moment capacity $M_D = 8.3\,\text{MN m}$. The M–N interaction diagram is constructed (Fig. 9.7(b)), with the applied forces superimposed. The applied loads are within the curve, so the section is adequate for these loads.

Fabrication of curved sections

Curved steel elements can be of two types, a true curve or a faceted curve formed from a series of straight elements. The use of a faceted curve has structural advantages, as it is relatively simple to fabricate and avoids the additional curvature moments between nodes. The truly curved section is, however, often preferred for aesthetic reasons. The bending of steel sections can be carried out by rolling or by induction bending [103].

Rolling of a steel section into a curved shape is a cold bending process, usually carried out as a continuous, three-point bending by rollers. A number of passes through the rollers may be required to achieve the required curvature. The rolling causes a plastic deformation of the steel section and will leave residual stresses locked into the section; these stresses are about 20% of the steel yield stress but usually have little effect on the ultimate section capacity. Roller bending requires a considerable force, for larger tubular elements induction bending is preferable.

Induction bending is a hot bending process. The steel section is heated locally by an electric induction coil to 700–1000 °C, at the same time a force is applied from a fixed radius arm, deforming the heated section. The heat applied in the induction bending process means that the residual stresses will relax and be less than those induced by cold bending.

Nodes in tubular structures

The nodes for tubular structures can be complex, particularly where more than one plane of bracing is used. The intersections are of two primary types, gap joints

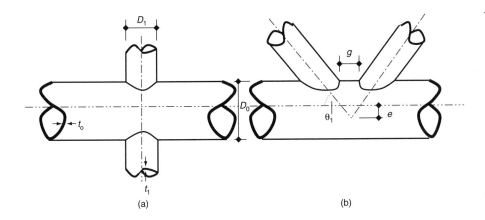

*Figure 9.9 Tubular node:
(a) X joint; (b) K joint.*

(Fig. 9.9(a)) or overlap joints (Fig. 9.9(b)). The strength of the node is greater for the overlap type. Other factors such as the brace angle, the relative size of the chord to the bracing and the thickness of the tubes also influence the node strength.

Rules for the design of nodes between tubular sections have been developed [104]. For the K and X joints three failure modes could occur: yielding of the brace member (N_y), plastification of the chord (N_p) or punching failure of the joint (N_v). The joint capacity may be estimated from the following equations:

$$N_y = 0.95 f_y A_{a1} \tag{9.8a}$$

$$N_{px} = k_2 f_y t_o^2 \left(\frac{5.2}{1 - 0.81\beta} \right) \tag{9.8b}$$

$$N_{pk} = \frac{f_y t_o^2}{\sin \theta} (1.8 + 10.2\beta) k_2 k_3 \tag{9.8c}$$

$$N_v = \pi v_y t_o D_1 \left(\frac{1 + \sin \theta}{2 \sin^2 \theta} \right) \tag{9.8d}$$

where A_{a1} is the brace area, k_2 is a coefficient dependent on the load ratio of the main chord (Fig. 9.10(a)), k_3 is a coefficient dependent on the overlap or gap in the K joint (Fig. 9.10(b)), β is the ratio of brace to chord diameters (D_1/D_0), v_y is the limiting shear stress in the steel (Equation (1.10)) and D_0, D_1 and t_o and so on are shown in Fig. 9.9. Where multi-planar joints occur the joint capacity should be reduced to 90% of the calculated X joint, unless more detailed interaction is considered [104].

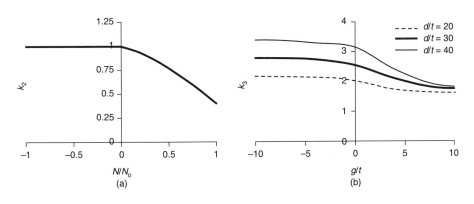

Figure 9.10 Tubular joint capacity coefficients: (a) k_2; (b) k_3.

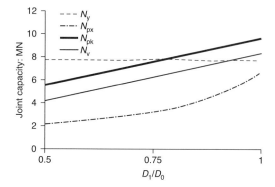

Figure 9.11 Node capacity for Example 9.2.

For the large tubes found in bridges, profiling of the tube ends at joints will be required to achieve welding fit-up. Joints can be strengthened by addition of external plates or internal ring stiffening. For composite structures the joint will be strengthened by the concrete and the punching capacity increased. The joint may be designed neglecting the contribution of the concrete and assuming the force from the brace is initially transferred to the steel section. Where the force transfer is large the local interface shear may be significant and connectors may be required between the steel and concrete. Typically connectors are required if the shear stress at the interface of tube and concrete exceeds $0.4\,\text{N/mm}^2$.

For the tubes of Example 9.2 the capacity of a node is calculated using Equations (9.8a to 9.8d), for various brace diameters. The results are shown in Fig. 9.11; for this example the brace yield and punching failures are not critical and the capacity is governed by chord plastification of the X joint for a wide range of relative tube diameters and thicknesses. For this example the node capacity can be significantly enhanced by the use of ring stiffening within the tube.

Aesthetics

The way bridges look, their character, is intimately bound up with the technical decisions taken. The earliest rules of architecture [78] list the requirements of stability, quality, amenity and economy, more a guide to the engineering qualities than a check list for aesthetics. Even in the late nineteenth century the architectural maxim of 'form follows function' was stated [1]. For early practitioners the rules of structure, aesthetics and practical construction were interchangeable. The rules of proportion were used to ensure the structural elements were of the correct thickness for stability as much as visual quality. Throughout the twentieth century, with greater specialisation, this holistic approach has been partly lost.

Many factors may affect the aesthetic decisions; these may be political, environmental, historic, technical, even current fashions. The political influence is often paramount, bridges form part of the larger road or rail infrastructure. The need to develop the infrastructure for the benefit of the economy of a country or area will be a political decision. The directives issued will dictate the priorities of the project, for many schemes speed of delivery or cost will be key requirements, occasionally political patronage will demand a statement or landmark structure, where looks or making an impression are key requirements. The environment in which a bridge is located is another key factor. Structures located in a beautiful landscape will have a head start in aesthetic merit over those sited in a decaying

Figure 9.12 Perspective sketch of a fashionable leaning arch structure across a motorway.

industrial environment or spanning a motorway. This does not mean that a structure crossing a great river will be beautiful (see the quotation in the Foreword), or that an attempt at some aesthetic merit cannot be made for motorway structures (Fig. 9.12).

The history of a site may influence aesthetics, particularly if there has been a bridge there before. The technical considerations are many, the length of span will affect the form and character (Fig. 1.1). The load carried will limit the aesthetic choice. Railway bridges with their heavy loads and tight deflection limits will develop along different lines from say footbridges, with their far lower loads. Fashion, or the prevailing taste swirling through society, has its effects and influences the bridge designer. The ornamentation, or lack of it, will be dictated by the prevailing styles. The ebb and flow of fashion in bridges is not as dramatic as the changes to hemlines on the Paris catwalk but its influence exists. In the last decade of the twentieth century and into the twenty-first the fashion is to lean arches or towers for visual reasons (Figs 9.12 and 10.10), to distort the structure from the traditional. Some of these structures will define an era or style, others will date relatively quickly.

The aesthetics of a bridge are governed by visual rules; these criteria are difficult to quantify in the same way as rules for producing a safe or serviceable structure, they are related as much to art as science, 'sculpting bridges' [79] perhaps. The rules are open to individual interpretation, the importance of any of the various parameters will depend on the individual designer and the prevailing influences.

Unity

A bridge should have a visual unity, a feeling of wholeness; 'some configurations are dominated by wholeness, others tend to separate' [107]. This property is often found in relatively simple arrangements, in structures with structural continuity, where the elements are clearly seen to be connected and where the superstructure and the supporting substructure seem related. Without unity the bridge will look like an assemblage of parts. The form of the bridge, the distribution of the visual masses and the degree of symmetry are important parameters to achieving this wholeness.

Scale

The bridge will be perceived with a size; if appropriate it will seem in scale. This does not mean the structure has to appear small. The size, the bigness of a bridge, may be appropriate, the dominant scale giving a sense of safety, that it has the strength to span. The bridge should be appropriate at different scales, when viewed close up its scale should not appear oppressive.

Proportion

This is the relationship of the parts to each other and to the whole. Geometric ratios, the span-to-depth ratio, the ratio of parapet to beam, the relative size of a span on a continuous bridge or the solidity ratio of the bays in a truss may all be used to give some proportion.

Rhythm

This is the pattern of repetition of elements, the order in which elements are placed, the rate of change in proportion. For bridges this is usually a horizontal rhythm, from many views the bridge will have a linear arrangement.

Detail

Detail is how the elements are shaped, and brought together with any ornamentation. Poor detailing, or an absence of detail will affect the wholeness, scale, proportion and rhythm of the bridge. It will affect the play of light and shade, its perceived colour and how it weathers.

To assess the many visual properties of a design, the bridge should be considered from a number of viewpoints; distant views, the middle distance and close up. All the criteria need to be assessed from all viewpoints; however, as in most design some combinations of properties and viewpoints are more critical. Scale and wholeness are more critical for distant views, detail is crucial for close views, rhythm and proportion being perceived more clearly at the middle distance.

At Runnymede the influence of aesthetics on the design has been significant for both structures. In the 1930s when the Staines bypass was proposed the Thames at this location was more rural than it is now. When it was built in the 1960s there was a political push to develop the nation's infrastructure. By the 1980s with the extensive development of the area, the M25 motorway (incorporating the original bridge) was heavily congested, and again there was a political push to widen congested motorways (see Chapter 4). Runnymede has now largely lost its rural feel, yet there is still a sense of calm at river level with light playing on the white soffit, despite the roar of the motorway above.

The brick-faced single span across the river borrows from the nearby 'great leap' double arch. The modern tubular form is influenced by the early tubular bridges of Eads and Baker, which [108, 109] span other great rivers. Its layout derived from a hand, as a series of fingers, the waiter's fingers in architectural terms [110]. There could have been other influences; a flying bird, a bull's head or a butterfly have all influenced the styling of other bridges [111]. Both bridges had architectural input, another cyclic fashion that ebbs and flows with bridge engineers' interest and training in aesthetics.

Tied arches

Where the abutments cannot carry arch thrusts a tied arch form may be appropriate, whereby the arch thrusts are tied through the deck. The large tensions induced in the tie means these structures are usually steel, although prestressed concrete ties have been used. The tied arch comes in three basic forms, with a dominant arch [113], dominant beams [114] or similarly sized arch and beam [115]. Visually each has its own character with different visual proportions. Structurally the bridges also have their own character each behaving differently. With the beam dominant, the arch acts only as the compression chord and carries permanent or uniform loads, its bending stiffness is significantly less than the deck beam stiffness so almost all moments are attracted to and carried by the deck beam. Where the arch is dominant it carries all moments and compressive forces leaving the deck to carry only the tie force and any local moments from the span between the vertical hangars; in this form the arch is often a large steel truss. Where arch and deck are of similar stiffness both share the bending effects of asymmetric loads.

For all tied arches the connection of arch to deck is a critical area, the compression of the arch and the tension of the tie must be resolved, and the vertical components must also be transferred to the substructure and foundations, the hinged arch clarifies this resolution structurally to some degree [112]. Often the arch and deck may not be in the same plane and the transfer needs to carefully consider the forces. The connection used will also significantly affect the character of the structure, and must be detailed carefully.

Example 9.3

The final example considered in this chapter is a design for a composite bowstring arch bridge spanning a river. The form is a continuous, twin girder ladder beam 330 m long with a central 187 m arch span. The structure uses composite action on the deck with a conventional ladder beam and slab layout. The arch compression chord is also a composite section formed from a fabricated steel U and box section filled with concrete. The deck of the bridge is continuous through the arch and approach viaducts, avoiding a joint and heavy abutment at this location. The continuity improves the durability of the structure, it ties the arch and approaches visually, and it also has significant structural benefits.

The continuity of the deck beams gives a stiffer structure and results in less moment in the arch. The continuity also helps spread the concentrated arch thrust, the placing of the arch tie inboard of the deck edge also assists with this spread. Figure 9.13 shows the shear stress contours in the concrete slab of a girder-stiffened

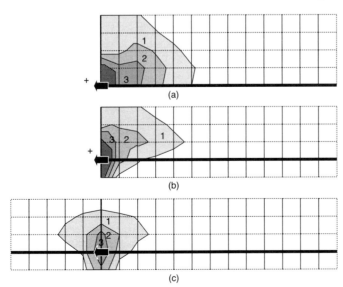

Figure 9.13 Shear stress concentrations from in-plane tie force from the arch: (a) non-continuous deck with edge beam; (b) non-continuous deck with inset beam; (c) continuous deck with inset beam.

bridge deck, with a concentrated in-plane force applied from the arch thrust. At midspan all stresses in the deck are relatively uniform with 20–25% carried by the deck and 75–80% by the steel girder. Where the arch and tie beam are at the edge of the slab the shear stress concentration is largest (Fig. 9.13(a)) [136]. Where the arch and tie beam are inset the stress magnitudes are similar but occur over a reduced area (Fig. 9.13(b)). For the bridge with continuity and an inset arch, the magnitudes of the stress concentrations are significantly reduced (Fig. 9.13(c)).

Arch buckling

Where the arch is part of a rigid frame the effective lengths of the arch members will be similar to those of a truss (Table 1.4). Where the arch is not restrained in its plane and is free to deflect then the effective length will be determined by the degree of restraint at the springing point and to some degree the tie and hanger stiffness. Figure 9.14 outlines some approximate effective lengths for in-plane buckling where the arch is the dominant form. In the transverse direction where the arch is unrestrained by a bracing frame, U-frame restraint stiffness K_H may be calculated (as Chapter 4). Where hangers with no bending stiffness are used then an effective restraint may be estimated from the horizontal force component (F) of the hanger load (N_H) as the arch deflects (δ).

$$K_H = \frac{F}{\delta} = \frac{N_H}{L_H} \tag{9.9}$$

Loading and analysis

The weight of the main span structure is 100 MN with surfacing. Maximum characteristic live loads are 40 MN; for the arch and deck design other load cases with more concentrated asymmetric loads will be more critical (as has been the case for all the arch examples). For a structure of this kind a number of frame models and submodels of various elements or connections are required, particularly since the arches lean slightly and moments will occur on both primary arch axes. The results of the analysis of an arch frame are shown in Fig. 9.15 for the serviceability limit state, indicating the distribution of moment, axial force and shear in the bridge. Table 9.2 tabulates the design values at the ultimate and serviceability limit state.

As with all composite structures the distribution of stresses is to a large degree dependent on the construction sequence. For a bridge of this size the sequence is

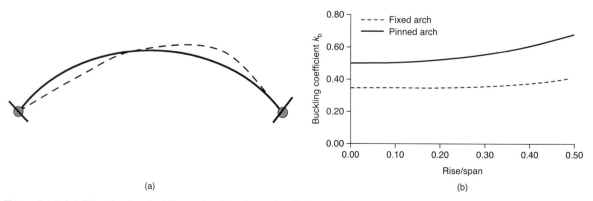

(a)

(b)

Figure 9.14 (a) First in-plane buckling mode; (b) effective length for in-plane arch buckling for a pinned and fixed arch.

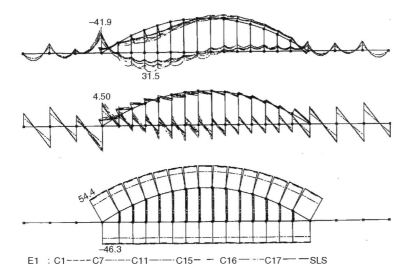

E1 : C1---- C7 --·-- C11 --·--- C15 -- · -- C16 -- ·-- C17 --- SLS

Figure 9.15 Tied-arch analysis results for the serviceability limit state.

relatively complex involving launching of the deck steel and propped construction of both the tie beam and arch. A simplified sequence is shown in Fig. 9.16. Due to the propping, the whole self-weight and live load are assumed to be carried on the composite structure.

For the arch the axial load capacity and contribution factors are calculated using Equations (9.2) and (9.3) as before. For this example:

$$N_{ul} = 0.56 f_{ck} A_c + 0.95 f_y A_a + 0.87 f_{ys} A_s = 122 \, \text{MN}$$

$$\alpha_c = \frac{0.45 f_{ck} A_c}{N_{ul}} = 0.6$$

indicating that a significant proportion of the capacity is derived from the concrete. In the transverse direction, bracing struts are provided at 40 m centres, giving an effective length of 34 m. In-plane the effective length is estimated from Fig. 9.14, conservatively neglecting the restraint from the hangars and tie. For an h/L ratio of 0.2 the effective length factor is 0.35.

$$L_e = 187 \times 0.35 = 65 \, \text{m}$$

Using Equation (9.6):

$$N_{cr} = \frac{(EI)_{a-c}}{L_e^2} \pi^2 = \frac{210\,000 \times \pi^2}{65^2} = 490 \, \text{MN}$$

Table 9.2 Tied-arch analysis results for design of the arch, tie and arch–tie intersection

Member	SLS			ULS		
	M: MN m	V: MN	N: MN	M: MN m	V: MN	N: MN
Arch at springing	6.0	1.0	54			
Tie at springing	4.1	4.5	46			
Arch at quarter				−24	1.0	61
Tie at quarter				−28	3.1	42

$$\lambda' = \left(\frac{N}{N_{cr}}\right)^{0.5} = 0.36, \text{ from Fig. 9.5, } k_1 = 0.95$$

$$N_D = 0.85 k_1 N_{ul} = 0.85 \times 0.95 \times 122 = 99 \text{ MN}$$

The axial capacity of the arch is well above the actual forces in the arch, an interaction diagram (similar to Fig. 9.7) can be drawn to show that the axial–bending interaction is satisfactory.

For the tie at the quarter point, the composite girder and slab section will carry a tension and moment. The section properties are given in Appendix C, the ultimate and serviceability stresses can be calculated. At the ultimate limit state, the tension plus sagging moment is critical for the lower flange. At serviceability, the stress in

Stage 1:

Stage 2:

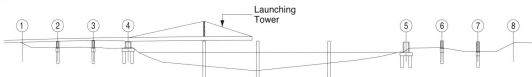

Stage 3:

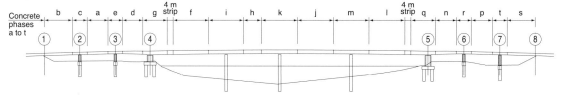

Stage 4:

Stage 5:

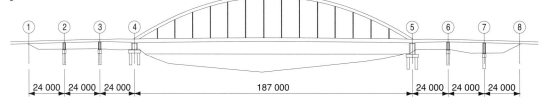

Figure 9.16 Construction sequence for the tied-arch bridge of Example 9.3: (a) construct substructure and temporary foundations in river; (b) launch steel deck; (c) concrete deck in stages as shown, leaving strips adjacent to the springing; (d) construct steel arch on temporary towers, concrete arch; (e) tension hangars, remove temporary support from river, concrete slab strips, place finishes.

the slab is critical, with tension plus a hogging moment. At 125 N/mm² the stress in the steel is low enough to prevent extensive cracking, provided the bars are at 150 mm centres or less (see Chapter 4).

At the arch tie intersection, the forces from the arch, tie, continuity beam and pier are resolved. The neutral axis of the tie beam with its slab is relatively high, a large steel top flange is used to draw this up further, such that the primary force is resolved near the top flange without significant transfer to the web and lower flange, avoiding the need for diagonal stiffeners, side plates or split webs. Only the vertical force transferred to the pier (or future jacking point) requires stiffening. The arch tie joint is shown in Fig. 9.17, with the primary forces.

At the arch tie intersection the force from the concrete arch section has to be transferred to the steel tie. The transfer, via connectors, could occur in the arch such that at the arch tie intersection all the force is in the steel. From the contribution factor, 60% of the force is in the concrete and the steel arch section must be thickened considerably to carry this additional load. Alternatively the transfer could occur at the tie beam connection, with connectors placed on the tie beam to carry the shear at this interface, a large number of connectors will be required. A third option is to use a combination of the two transfer methods, for this example this method will be used.

At the serviceability limit state, the arch at the intersection carries 32 MN in the concrete and 22 MN in the steel. This is modified to 20 MN in the concrete with 34 MN in the steel by some transfer of load across connectors on the arch. To carry the additional load the top and bottom flanges of the box are doubled in size (the webs are not increased as additional loads here are more difficult to resolve without using a box section tie beam local to the intersection). Assuming a 4 m length of transfer (based on a 1:2 load spread as Chapter 3), the force to be transferred is:

$$Q_1 = \frac{(32 - 20)}{(2 \times 4)} = 1.5 \, \text{MN/m}$$

Using Equation (1.17) with 22 mm-diameter connectors, with $P_u = 139$ kN (Table 1.5):

$$No = \frac{Q_1}{0.55 P_u} = \frac{1.5}{0.55 \times 0.139} = 20$$

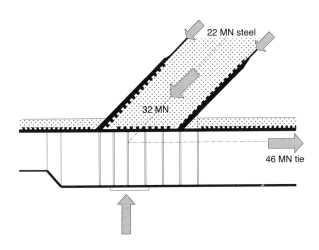

22 MN steel

32 MN

46 MN tie

Figure 9.17 Tie intersection for Example 9.3 showing serviceability limit state forces.

Six 22 mm-diameter connectors in rows at 300 mm centres over the last 4 m of the arch flanges will be adequate.

At the steel–concrete intersection on the tie beam the remaining 20 MN in the concrete will require connectors to carry the horizontal component of the force. The force is assumed to be carried on a 2.5 m length of beam. Resolving forces:

$$Q_1 = \frac{20 \times 47}{2.5 \times 54} = 7\,\text{MN/m}$$

This is a relatively large force and bar connectors will be used, $P_u = 963\,\text{kN}$ (Table 1.5):

$$No = \frac{Q_1}{0.55 P_u} = \frac{7.0}{0.55 \times 0.963} = 13$$

Ten rows of four $50 \times 40 \times 200$ mm bars with hoops will carry this load on the top of the tie within the arch. For both shear planes reinforcement will be required within the arch across the internal shear planes.

10

Cable-stayed bridges

...have a system of forces that are resolved within the deck–stay–tower system...

Introduction

The development of cable-stayed bridges was contemporary with that of composite structures; both arose primarily in the later part of the twentieth century. All of the long-span cable-stayed bridges combine steel and concrete elements; many have some composite components [20, 116, 117]. A cable-stayed bridge consists of a deck stayed by high-tensile wire or strand from towers. The most common form is a single main span with smaller back spans and two towers (see Fig. 10.2). Asymmetric layouts hung from a single tower are popular for smaller spans [118] and architectural statements [111]. Multi-span cable-stayed structures are now also being developed [119] with more rigid frame towers. The stay cables are arranged in a harp or semi-fan arrangement [57], the true fan arrangement, with all stays converging at one point, is difficult to detail and rarely used.

Cable-stayed bridges tend to be self-anchored, that is they have a system of forces that are resolved within the deck–stay–tower system. Earth-anchored systems utilising massive anchorages (similar to suspension bridges) to increase the system stiffness have been proposed as a way of increasing spans [120].

The basic stay system is fundamentally a series of superimposed triangular trusses. An approximation of the behaviour can be obtained relatively simply; however, the bending, shear and axial load interaction together with the non-linear behaviour of the stays can make detailed analysis relatively complex. Consider an isolated deck–stay–tower system as that shown in Fig. 10.1. Element 1 of the main span has a weight W_1 and is located a distance L_1 from the tower, it is attached to a tower of height h_1. A tension T_1 in the stay and compression C_1 in the deck are required for stability.

$$C_1 = W_1 L_1 / h \qquad\qquad (10.1a)$$
$$C_2 = W_2 L_2 / h \qquad\qquad (10.1b)$$

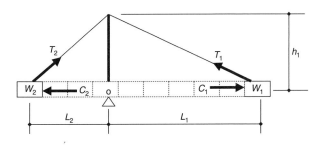

Figure 10.1 Forces in a stay system.

$$T_1 = \left(W_1^2 + C_1^2\right)^{0.5} \tag{10.2a}$$

$$T_2 = \left(W_2^2 + C_2^2\right)^{0.5} \tag{10.2b}$$

To avoid out-of-balance forces at the tower top and in the deck, $C_1 = C_2$, and

$$W_2 = W_1 L_1 / L_2 \tag{10.3}$$

which also gives equilibrium about point o of Fig. 10.1. Hence, if a large span is being planned then either the back span has to be approximately half the length of the main span or it needs to be significantly heavier. This leads to the two primary layouts of bridges with stiffening girders and relatively long back spans or bridges with short, heavy back spans supported on multiple piers. For a 1000 m ($L_1 = 500$ m) span and a 250 m-long back span (L_2) then the deck in this section would be at least twice as heavy as the main span. If a lighter steel main span is used then heavier, concrete back spans seem logical (Fig. 10.2). If a live load is placed on element W_1 then element W_2 needs to be tied down or given sufficient bending stiffness to span to the supports.

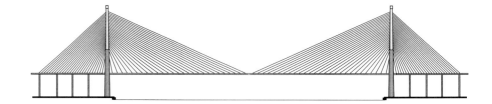

Figure 10.2 Stonecutters Bridge: light, steel span (1018 m) with heavy concrete back spans (258 m).

Deck–stay connection

Composite cable-stayed bridges utilise the best characteristics of the two materials, the vertical component of the stay being resisted by the steel girder and the horizontal component mainly by the concrete deck slab. Ideally the cable connection should permit this.

The connection between the stay and deck can take a number of forms (Fig. 10.3). The stays may be anchored into a large, concrete edge beam replacing the longitudinal steel girder, and leaving only the composite cross-beams, the longitudinal and vertical forces being carried entirely by only the concrete section. If the

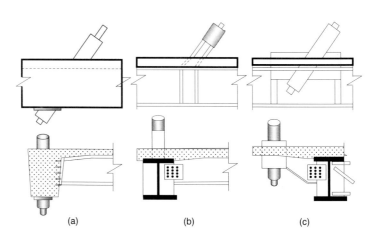

Figure 10.3 Stay–deck connections: (a) Tagus Bridge; (b) the Boyne Bridge; (c) Second Severn Bridge.

(a)   (b)   (c)

stay is attached directly to the steel girder, all loads are initially carried by the steel-work. The horizontal component is then largely transferred to the concrete deck that forms the majority of the section's area via shear connectors. For the Second Severn Bridge the connection is via a transfer structure that resolves horizontal loads directly to the deck slab and the vertical components to the main girders [116].

Example 10.1

The first example of a composite cable-stayed bridge is the Second Severn Bridge, a 456 m-span, 34 m-wide structure spanning the shoots channel of the River Severn, 5 km downstream of the first crossing connecting England and Wales. The bridge has a twin stay plane, each with 120 cables in a semi-fan arrangement. The length of the back span is just under half of the main span giving a 900 m-long cable-suspended bridge. A stiffening girder is used to resist live load moments and the back spans are further stiffened with intermediate piers. The deck stiffening consists of two longitudinal steel plate girders 2.15 m-deep with transverse composite trusses at 3.6 m centres supporting the deck slab. The slab thickness varies between 350 mm at the girders, 470 mm at the cable anchorage and 200 mm for the majority of the width between girders. The deck concrete has a strength of $55/70 \, \text{N/mm}^2$.

High-strength concrete

Typical concrete strengths are within the 25/30 to $40/50 \, \text{N/mm}^2$ range; this range is also used in current design codes [4, 7]. With careful consideration of the aggregate type, a reduction in the water–cement ratio and some cement replacement and additive, concrete strengths of up to $92/110 \, \text{N/mm}^2$ are possible. For steel–concrete composite structures the use of high-strength concrete will allow lower volumes of concrete and lighter structures. The higher-strength concrete will also have a higher elastic modulus and lower shrinkage and creep values, leading to an increase in stiffness [121].

For sections with high-strength concrete in compression the design of the section will be similar to conventional-strength sections (Equation (1.4)). The limiting strain used in design will reduce with strength. At high strengths this may lead to a loss of ductility (see Fig. 1.2):

$$\varepsilon_u = 0.0035 - \frac{(f_{cu} - 60)}{50\,000} \tag{10.4}$$

The tensile strength of the concrete will also increase, and Equation (1.1) is modified:

$$f_{ct} = 0.58(f_{ck})^{0.5} \tag{10.5}$$

The increased tensile strength will increase the load at which first cracking occurs in continuous beam and slab construction at supports. The reinforcement in the tensile zone must be sufficient to limit crack widths, the increased tensile capacity leading to an increase in the minimum reinforcement. For concrete strengths over $45/55 \, \text{N/mm}^2$ the minimum reinforcement should be:

$$100A_s = 0.13\left(\frac{f_{cu}}{40}\right)^{0.67} \tag{10.6}$$

The lower water–cement ratios of high-strength concrete (typically below 0.3) mean that shrinkage and creep are usually reduced by 20 to 50%. The stiffness of high-strength concrete elements is also increased, and the elastic modulus will increase (as Table 1.1).

Loads

For the suspended span the dead loads, particularly that of the slab, should be minimised to reduce the volume of the relatively expensive stay cables. A 200 mm-thick slab is used where possible, thickening at the edges where high local shear stresses occur near the stay anchorages.

Highway loads

The live load for bridges [6, 31, 32] varies with span, for longer span lengths the load intensity is generally reduced (Fig. 10.4). The loading on a length of highway is dependent upon the vehicle weight, its length, axle configuration and the number of vehicles bunched together. For bridges of short span the impact and overloading factors are also important. For longer loaded lengths the likelihood of a series of fully loaded heavy vehicles travelling across the structure is reduced; it is also likely that some of the vehicles will not be fully loaded. The proportion of heavy vehicles depends upon the prevailing economic conditions of the country or area in which the bridge is located. This will vary with time and may be influenced by the bridge itself.

The loading used in the UK has increased. Figure 10.4 shows the HA loading used for the design of the original Severn Bridge from 1966 [122] and HA loading used in 2004 for the Second Severn Bridge. The number of loaded lanes has also increased from a full lane with adjacent lanes having one third full configuration to a full lane with two thirds full in adjacent lanes, leading to an effective doubling of total load on the highway.

Wind loads

For long-span suspended structures, wind loading will govern aspects of the design and needs to be carefully considered. The basic wind speed at the site is determined from data in standards [21, 32] but can be derived from measured values at the site if necessary. From this basic site hourly wind speed (V_s) the design wind speed (V_d) taking into account the terrain, height, direction and

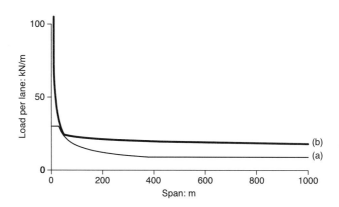

Figure 10.4 Live load variation with length: (a) HA loads circa 1966; (b) HA loads for 2004.

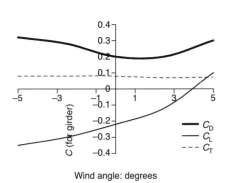
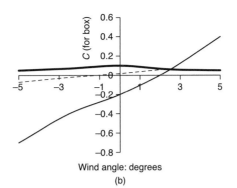

Figure 10.5 Variation of drag, lift, twist with wind angle: (a) girder bridge; (b) box shape.

size of structure can be determined:

$$V_d = V_s k_d k_a k_p k_g \tag{10.7}$$

where k_d is a factor for the wind direction, k_a depends upon the altitude, k_p is a probability factor and k_g is a gust factor depending on the terrain, structure height and size. From the velocity the wind pressure (q) can be determined:

$$q = 0.613 C V_d^2 \tag{10.8}$$

where C is a drag coefficient. The wind imposes load on the structure causing drag (C_D), lift (C_L) and twisting (C_T) of the suspended deck. Figure 10.5 shows the typical variation in these coefficients with the angle of the wind incidence, for a relatively bluff girder bridge and a more aerodynamic box section. The magnitudes of the coefficients are also affected by the shape and width of the deck.

This static loading is only part of the wind effect. Wind velocities vary with time and gusting causes dynamic effects such as buffeting, flutter and the shedding of vortices. The susceptibility of a bridge to dynamic wind effects can be determined by a factor P [123]:

$$P = \left(\frac{\rho b^2}{m}\right)\left(\frac{16 V_s^2}{b L f_b^2}\right) \tag{10.9}$$

where ρ is the density of air, b is the bridge width, L is the span, m is the mass per unit length of the bridge and f_b is the first bending frequency of the structure. For $P < 0.04$ the structure is unlikely to be susceptible to aerodynamic excitation. For $P < 1$ the structure should be checked against some simplified criteria to check for any aerodynamic instability. If $P > 1$ the structure is likely to be susceptible to aerodynamic excitation and some changes to the mass, stiffness or structure layout may be required, and wind tunnel testing will be required to verify the structure's behaviour.

For the Second Severn Bridge example, $b = 35$ m, $L = 456$ m, $m = 40\,000$ kg/m, $V_s = 37$ m/s. The bending frequencies of cable-stayed bridges tend to be lower than other beam, arch or truss type structures (Fig. 4.10). For this example $f_b = 0.33$ Hz [124]. Using Equation (10.9):

$$P = \left(\frac{\rho b^2}{m}\right)\left(\frac{16 V_s^2}{b L f_b^2}\right) = \frac{(1.25 \times 35^2)}{40\,000} \times \frac{(16 \times 37^2)}{(35 \times 456 \times 0.33^2)}$$

$$= 0.036 \times 13 = 0.47$$

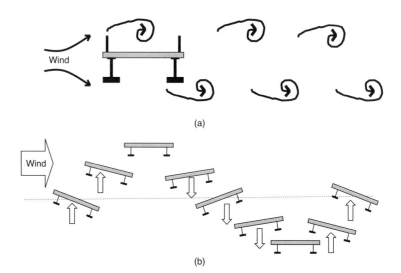

*Figure 10.6 (a) Vortex
shedding from a bridge
causing movement of the
structure; (b) flutter of a
bridge involving combined
vertical and torsional
movement.*

In this case $0.04 < P < 1.0$, so some simplified checks on two principal causes of wind excitation, vortex shedding and flutter will be carried out.

Vortex excitation is caused by the shedding of vortices in phase with the natural frequency of the structure (Fig. 10.6(a)). The critical wind speed at which this occurs can be established from:

$$V_{cv} = \frac{f_b d}{S} \qquad (10.10)$$

where d is the depth of the structure and S is the Strouhal number, this varies from approximately 0.12 for a bluff girder system to 0.3 for a smooth cylinder shape.

For the example the depth of the structure is approximately 3.1 m; however, wind shielding 3 m high is provided at the edge of the bridge to lower the wind loading on traffic. Wind tunnel testing on the barriers [94, 116] indicated they have some effect on the aerodynamic behaviour of the deck. Using Equation (10.10) with $d = 3.1$ m and $S = 0.12$:

$$V_{cv} = \frac{0.33 \times 3.1}{0.12} = 8.5\,\text{m/s}$$

If a larger depth is used to allow for wind shielding:

$$V_{cr} = \frac{0.33 \times 6.1}{0.12} \quad \text{giving } 16.5\,\text{m/s}$$

For high bridges the wind may not always be horizontal, and this may increase the effective depth. For the example, the deck is 40 m above the water and a $3°$ angle of attack (θ) seems a reasonable assumption, and $d = d + L\sin\theta = 6.1 + 1.8 = 7.9$ m. Recalculating the critical wind speed:

$$V_{cv} = \frac{0.33 \times 7.9}{0.12} = 21.8\,\text{m/s}$$

A large range of critical velocities is calculated reflecting the various assumptions. All the velocities are lower than the hourly site wind speed and may occur relatively frequently, the higher value is a significant proportion of the site wind speed and may have a significant force input as the wind force is proportional to the square of the velocity (Equation (10.8)).

Flutter is an aerodynamic phenomenon involving a combined vertical and torsional movement of the bridge deck (Fig. 10.6(b)). The critical wind speed at which this flutter occurs on a thin plate can be estimated from:

$$V_{cf} = k_f V_{fd} \left[1 - \left(\frac{f_b}{f_t} \right)^2 \right]^{0.5} \tag{10.11}$$

For bridge deck sections k_f varies from 0.4 for bluff sections to 0.9 for more aerodynamic shapes, f_t is the first torsional frequency and V_{fd} is the divergent wind speed:

$$V_{fd} = (f_b / r_m) \left(\frac{\pi J_m}{\rho} \right)^{0.5} \tag{10.12}$$

where J_m is the mass moment of inertia $(I_p m / A)$ and r_m the mass radius of inertia $(J_m / m)^{0.5}$, and I_p is the polar moment of inertia $(Ix + Iy)$.

For this example $I_p = 88 \, \mathrm{m}^4$, $J_m = 4.8 \times 10^6 \, \mathrm{kg/m^2}$ per m, $r_m = 10.9 \, \mathrm{m}$. The torsional frequency for the bridge is 0.47 Hz and using Equation (10.12):

$$V_{fd} = \left(\frac{f_b}{r_m} \right) \left(\frac{\pi J_m}{\rho} \right)^{0.5} = \left(\frac{0.47}{10.9} \right) \times \left(\frac{3.14 \times 4.8 \times 10^6}{1.25} \right)^{0.5} = 149 \, \mathrm{m/s}$$

And using Equation (10.11) with $k_f = 0.45$

$$V_{cf} = k_f V_{fd} \left[1 - \left(\frac{f_b}{f_t} \right)^2 \right]^{0.5} = 0.45 \times 149 \left[1 - \left(\frac{0.33}{0.47} \right)^2 \right]^{0.5} = 48 \, \mathrm{m/s}$$

This velocity is higher than the hourly wind speed but not by a significant margin. Given that the accuracy of the calculations above is limited, wind tunnel testing of the bridge cross-section is required to confirm the critical flutter velocities and to ensure any vortex excitation is of limited amplitude.

A series of wind tunnel tests were carried out on the structure [116, 124, 125]. The first tests on a 1:50-scale sectional model looked at the effect of removing the enclosure shown on the original design. Removal of the enclosure affects the drag (similarly to Fig. 10.5), but also changes the flutter and vortex response, increasing V_{cf} and showing some vortex shedding. A second series of sectional models were then carried out on the section with the primary steel stiffening girders in various locations. It was found that layouts with girders near the edge (which are structurally more efficient) caused more vortex shedding than those with girders nearer the centre of the section. The final deck girder layout with the girders inset from the edges was a compromise between conflicting structural and aerodynamic requirements. The testing of this section showed some limited vortex-induced response under smooth flow, with none where turbulent flow was used. A final set of testing on a 1:125-scale model of the bridge (stays, deck and towers) was carried out. This test showed vortex shedding in smooth flow at a wind inclination of 2.5°; at horizontal inclinations or with turbulent flow no vortex shedding response was recorded.

In the first winter of operation the bridge exhibited some vertical oscillation. The structure was instrumented and the response to the wind measured. After analysing the wind and the structure's response to it, a further series of wind tunnel tests were

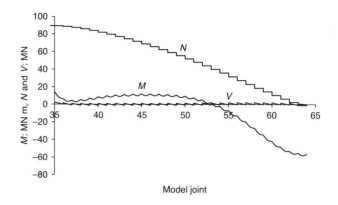

Figure 10.7 Analysis results for Example 10.1: (a) moment; (b) shear; (c) axial load for live load.

carried out to replicate the measured response. Based on these studies it was concluded that both the structural damping and the wind turbulence were lower than assumed for design. These conclusions are relevant to many other bridges of this type [124]. It is also clear that changing the cross-section has a significant effect on the design. A further set of wind tunnel tests were then made with modifications to the structure. The modifications consisted of two longitudinal baffle plates between the girders for approximately one third of the span, the tests indicated that this would suppress vortex oscillations. The baffles were fitted to the structure and measurements indicate that vortex-induced oscillations have ceased.

Analysis

A first estimate of cable loads can be obtained from the structure's geometry. To estimate deflections, moments and shears in the girders and towers a more complex model with a high degree of redundancy is required. To understand the behaviour, a series of simple models are generally preferable to a large complex model. A two-dimensional linear elastic representation of the towers, piers, deck and cables with spring supports representing the foundations will usually be adequate. Depending on the stresses in the cables some modification of the material modulus may be required to allow for the cable sag [57], particularly for long-span structures.

During the life of the structure it is likely that the stays will need to be replaced, this will be carried out with traffic on the structure. It is also possible that one or more cables could fail, shedding load to adjacent cables; it is therefore important that the structure is robust and that a progressive 'unzipping' failure does not occur. For the cable-out scenarios a three-dimensional model of the bridge or a section of deck will usually be required. This model is also used to confirm the torsional frequencies of the bridge under wind loads.

Local models may also be required to represent the transfer of load from the stay to the deck, the use of a strut and tie model [140] will also give a good estimate of likely load paths. Figure 10.7 shows the global moment and shear of the deck for live loads based on the two-dimensional model with stays and decks joined at a node on the longitudinal girder. The stays are located at the edge of the slab and the girder is inboard of this, and the connection is via an anchorage beam and two cantilever beams. This spreads the load reducing the peak moments and shears on the girder.

Buckling interaction

For a slender steel–concrete composite section subject to shear, bending and axial loads buckling of the relatively slender components may occur. In Chapter 4 it was noted that there is an interaction between bending and shear, and in Chapters 8 and 9 the interaction of moments and axial load was explored. For a composite, cable-stayed bridge deck the girder system resists bending and shear with a large axial load, and there will be an interaction of all three [2]. The web of a girder in a cable-stayed bridge is particularly prone to this buckling, for a web panel the buckling interaction equation is:

$$m_c + m_b + 3m_v < 1 \tag{10.13}$$

$$m_c = \frac{0.95f_c}{k_c f_y}$$

$$m_b = \left(\frac{0.95f_b}{k_b f_y} \right)^2$$

$$m_v = \left(\frac{0.95v}{k_v f_y} \right)^2$$

and k_c, k_b and k_v are coefficients that vary with the web slenderness and degree of edge restraint (see Fig. 10.8).

For the bridge in Example 10.1 the design forces for the girder near the quarter point are obtained from Figure 10.7. At this location the girder is formed from a 20 mm-thick web and a 750 by 90 mm bottom flange. The section properties are obtained from Appendix C, the stresses at the ultimate limit state are:

$$f_{top} = \frac{N}{A} + \frac{M}{Z_t} = \frac{102}{0.535} + \frac{16}{1.47} = 190 + 11 = 201 \, \text{N/mm}^2$$

$$f_{bot} = \frac{N}{A} - \frac{M}{Z_b} = \frac{102}{0.535} - \frac{16}{0.21} = 190 - 76 = 114 \, \text{N/mm}^2$$

$$f_c = \frac{(f_{top} + f_{bot})}{2} = 158 \, \text{N/mm}^2$$

$$f_b = f_{top} - f_c = 44 \, \text{N/mm}^2$$

$$v = \frac{V}{dt} = \frac{2.7}{2.1 \times 0.020} = 64 \, \text{N/mm}^2$$

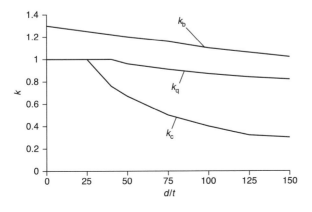

Figure 10.8 Buckling interaction coefficients for restrained and unrestrained panels.

*Figure 10.9 Cantilever
construction of the Second
Severn Bridge
(© T. Hambley).*

If the web is unstiffened, $d/t = 2100/20 = 105$. From Fig. 10.8, $k_b = 1.2$, $k_v = 0.94$ and $k_c = 0.4$.

$$m_b = \left(\frac{0.95 \times 44}{1.2 \times 355}\right)^2 = 0.01$$

$$m_v = \left(\frac{0.95 \times 64}{0.94 \times 355}\right)^2 = 0.03$$

$$m_c = \left(\frac{0.95 \times 158}{0.4 \times 355}\right) = 1.05$$

Thus $m_c + m_b + 3m_v = 1.15$; this is greater than 1 and so buckling may occur. It can be seen that for this example the axial buckling dominates the design. To increase the capacity of the section, stiffening is required, and for the built structure two longitudinal angle stiffeners were used dividing the web into three sections. For the stiffened web, $d/t = 35$, $k_b = 1.22$, $k_v = 1$, $k_c = 0.8$, $m_c = 0.63$: thus $m_c + m_b + 3m_v = 0.73$, which is now satisfactory. It is also possible to avoid buckling of the web by using a corrugated web (see Chapter 11).

Shear connection

In Chapter 1 it was noted that the force transferred across the steel–concrete interface is related to the rate of change of force in the slab (Equation (1.14)). For many bridges this can be simplified to the rate of change in bending – in proportion to the vertical shear (Equation (1.16)). For cable-stayed bridges where there is a significant axial load the rate of change in axial loads needs to be taken into account.

The rate of change in axial force can be visualised by considering the construction method for the bridge (Fig. 10.9). The bridge in this example is constructed using a cantilever technique with a 7.4 m segment of deck and stay added at each face. Each deck section is supported by a stay and puts a small increment of axial force into the deck. The force in this example is applied directly to the deck concrete (Fig. 10.3(c)). Part of this force will be carried by the deck steelwork, and shear connectors will be required to transfer the load across the interface.

$$\delta N = \frac{N A_a}{(A_a + A_c/n)} \tag{10.14}$$

$$Q_1 = \frac{\delta N}{L_s} \tag{10.15}$$

where L_s is the length over which the force is transferred. With stays at 7.4 m centres it is likely that using $L_s = 7.4$ m will give a reasonable distribution of connectors. The length over which the force is transferred depends upon the stiffness of the connectors and can be estimated from Equation (3.13).

For the Second Severn Bridge example, near the quarter point, $V = 2.0$ MN at the serviceability limit state and the increment in axial force between stays is 4.7 MN. From Appendix C, $Q_1/V = 0.44$, $A_a = 0.13\,\text{m}^2$, $A_c = 4.8\,\text{m}^2$ and the modular ratio is 12.

$$\delta N = 4.8 \times \frac{0.13}{(0.13 + (4.8/12))} = 1.18\,\text{MN}$$

$$Q_{1_N} = \frac{\delta N}{L_s} = \frac{1.18}{7.4} = 0.16\,\text{MN/m}$$

$$Q_{1_m} = 0.44 \times 2 = 0.88\,\text{MN/m}$$

The total longitudinal shear $Q_1 = 0.16 + 0.88 = 1.04\,\text{MN/m}$. Using Equation (1.17) with 25 mm connectors, $P_u = 183\,\text{kN}$ (Table 1.5 or Equation (1.19)):

$$No = \frac{1040}{0.55 \times 183} \text{ giving } 11$$

Two 25 mm-diameter connectors at 150 mm centres are satisfactory at this location.

Towers

The towers of a cable-stayed bridge provide much of its character. For a given span the taller tower of the asymmetric span will be more prominent than the smaller towers of the symmetric layout. Many of the monumental cable-stayed structures have exploited this asymmetric layout [100, 111, 127]. Towers are of three basic forms: the H shape, this is a simple and economic layout for medium-span bridges with two planes of cables (see Fig. 10.9 for the towers for the Second Severn bridge), the single leg tower and the inverted Y form [118]. The last two have the stays anchored in a line, either by the use of a single plane of cables, or by inclining cables on a twin plane of stays. The use of this inverted Y form gives some triangulation and increases the torsional stiffness of the deck–stay–tower system and is used for larger spans [20].

Figure 10.10 Leaning towers, front and rear views.

The three basic forms may be modified in many ways to achieve some architectural statement or visual interest. Single towers are often inclined in an attempt to drama-tise the structure, towers leaning back can achieve a certain sculptural dynamic from some views (Fig. 10.10), however, from other views the tower can appear to lean forwards leading to a more unstable perspective, particularly from the driver's view. The viewpoint for structures with a high degree of aesthetic input is often chosen to flatter and complement the design. When such structures are being judged, other viewpoints should normally be requested to avoid the possibility of a mediocre structure being selected on the basis of one viewpoint or image.

The leaning of towers and other elements of a structure usually make the struc-ture less efficient and more costly. Consider a simple, single-leg tower structure similar to Fig. 10.10. If the towers lean backwards or forward the relative volume of material in the tower and stays changes. Figure 10.11 shows the change in volume of materials for various inclinations, the steeper the lean the less efficient the structure becomes.

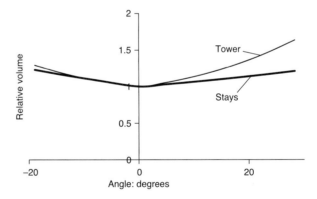

Figure 10.11 Variation in structure material volume with tower inclination.

Tower top

For the common semi-fan form of cable layout the tower can be divided into two sections, an upper section, in which the stays are anchored, and a lower section. The lower section carries the axial force to the foundations and provides the bending resistance to overturning (hence the usually tapered form, widening to the base). The forces in the upper tower section are more complex; this section has to resist smaller overall bending, shear and axial effects, but also has to resist the high local loads from the stay anchorage. For the majority of cable-stayed structures the stays are anchored to the sidewalls of the tower leading to a significant tension effect across the tower. For smaller structures [118], anchoring to the far side of the tower can reduce this complexity. For concrete towers the splitting can be resisted by transverse prestress [140]. The layout and anchoring of this prestress can be complex and many recent bridges [20, 126] have used a steel–concrete composite box arrangement to transfer this tensile load.

For the tower composite box structure, the inclined tension force of the stays is resolved horizontally and vertically. The horizontal component of the stay tension is resolved across the box, the transverse area of the box must be large enough to resist this tension. The vertical component of the stay causes compression on the box, which is carried locally by only the steel section. In a composite steel–concrete tower this force is then transferred via connectors to the concrete. If the vertical components on each side of the box are not the same, some bending will be induced in the section. Additional reinforcement in the concrete, or an external steel skin may be used to resist this moment.

Example 10.2

The transfer of forces in the top of a tower are complex due to the various angles of the stay anchorage; this complexity is explored in the second example of this chapter, the Stonecutters Bridge (Fig. 10.2). Stonecutters Bridge is a large 1018 m-span cable-stayed bridge across the Rambler Channel in Hong Kong, which is located between the Kwai Chung container port (formerly Stonecutters Island) and Tsing Yi Island. The bridge deck consists of twin box concrete girders in the side spans and twin box steel girders around the towers and in the main span. The deck boxes are separated with an air gap but are interconnected with transverse members at 18 m intervals; this arrangement significantly improves the flutter response of the structure [137]. It is a cable-stayed bridge with two tapered, cylindrical, single-leg towers, each 298 m tall (Fig. 10.12). The cables are arranged in a semi-fan configuration.

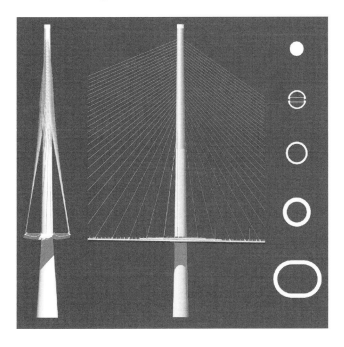

Figure 10.12 A tower of the Stonecutters Bridge with a composite upper section (© Arup).

The lower sections of the towers from top of pilecap to the 175 m level are formed of reinforced concrete. The upper sections to 293 m are of composite construction, with an internal high-strength steel box structure composite with a concrete surround and a stainless steel outer skin. The top 5 m of the towers are glazed and house lighting features. The upper tower tapers from 10.9 m diameter to 7 m diameter with a wall thickness varying from 1400 mm to 820 mm. The steel skin is a 20 mm-thick stainless steel plate and forms the outside face of the concrete tower. Composite action of the skin and the concrete wall is achieved by connecting the skin to the wall by means of shear studs. The design of the towers has developed from the original design [117], which proposed an all-steel top section. The steel–concrete composite section was used to increase stiffness and reduce costs.

Stainless steel

The twin properties of corrosion resistance combined with good aesthetics are the primary reasons for using stainless steel. Like weathering steel (Chapter 3) stainless steel forms a passive oxide film of corrosion products on its surface; this film limits further corrosion and means painting is not required. The oxide film can be damaged by abrasion, atmospheric pollution or a marine environment. There are a number of types of stainless steel. For structural use in bridges, austenitic and duplex stainless steels are the most common. Of these types the duplex is more tolerant of pollution and coastal environments.

The structural properties of stainless steel are similar to normal steel; the elastic modulus is slightly lower at $200 \, \text{kN/mm}^2$, the coefficient of thermal expansion is higher at 16×10^{-6} per °C. The stress–strain profile of stainless steel has a less pronounced yield point than normal steel plate (Fig. 1.4), and the strength is usually based on the 0.2% proof strength. For austenitic stainless steel the strength is typically 200 to $250 \, \text{N/mm}^2$, for duplex stainless steel the strength is 400 to $480 \, \text{N/mm}^2$.

Loads

The Stonecutters Bridge spans a major shipping channel and is located in a moderately seismically active area. The bridge is also located in an area subject to typhoon storm loads and so wind loads are also significant, adding transverse bending effects to the tower. These criteria are important for the lower tower and its foundations but less so for the upper tower. For the upper section, the dead and live loads transferred from the deck via the stays are the major influence.

Analysis

The design of the bridge has used a series of analyses at the various stages of design development from concept through to the detailed design phase. The final analysis consists of a three-dimensional frame analysis of the bridge (deck, stays, piers and towers) with stay non-linearity being taken into account. The tower consists of a tall structure with significant axial loads and second-order (P-Delta and critical buckling) effects [57] were also taken into account. The forces for the upper tower were then processed in a series of spreadsheets to extract maximum and minimum values of axial force and bending, together with their coexistent effects.

Considering a stay in the middle of the composite tower section, the reaction from the stays above this location is 280 MN. At this section the stay forces are 16 MN from the main span and 22 MN from the back span. Resolving the stay forces gives an applied local axial load of 7.0 MN and a horizontal component of 13 MN. The effective depth of the tie across the box is assumed to be 1.2 m with a thickness of 20 mm for each of the two sides. Using Equation (1.7): $N_{Dtie} = 0.95 f_y A_{ae} = 0.95 \times 420 \times 1.2 \times 0.04 = 19$ MN, more than the applied loads. A local three-dimensional finite element analysis of the steel skin, concrete tower and anchor box was also undertaken to confirm the local effects on the tower. The local and global effects are added together to determine the final stresses in the system.

The vertical component of the force will be transferred via connectors, the distance between stays is approximately 3 m and 70% of the local vertical component is reacted on the back span side of the box.

$$Q_l = 25 \times \frac{0.7}{3} = 5.8 \text{ MN/m}$$

Using Equation (1.17) and assuming 22 mm-diameter connectors ($P_u = 139$ kN from Table 1.5):

$$No = \frac{Q_l}{0.71 P_u} = \frac{5.8}{0.098} \text{ giving } 60$$

These connectors are spread across the box face in a staggered layout keeping the centres below 600 mm (Table 1.3).

The box and steel skin are formed of relatively thin (generally 20 mm-thick) steel plates, the concrete section and its reinforcement forming a larger area. The tower section was therefore designed using methods similar to those for reinforced concrete sections [140], where stresses are derived from strains. The steel skin contributes to both the vertical and hoop reinforcement. The stainless steel skin is formed in a series of 2 to 4 m tall sections with bolted splices. The splices are considered as compression-only elements, the tension capacity being conservatively neglected.

Strain-limited composite section

Steel–concrete composite compression members with a steel contribution factor (α_a) less than 0.2, or where the steel element thickness is outside the limits of Table 1.3, will be governed primarily by the behaviour of the concrete. The steel element will rely on connectors to prevent it buckling locally. The axial and bending capacities of the section will be limited by the strain distribution across the section, and not all of the section will be at yield (Fig. 10.13). The extreme tension and compression capacities of the section will be similar to a more compact or steel-governed section; the bending capacities of the section will be reduced.

For the Stonecutters tower the axial capacities of the composite section (N_{UL}) and the stainless steel skin (N_{ua}) are 760 MN and 225 MN, giving a steel contribution factor (α_a) of 0.3. An $M-N$ interaction curve for the tower can be constructed, and the applied moment and axial load are well within the section capacity. The shear at the interface of the stainless steel skin can be calculated; however, like

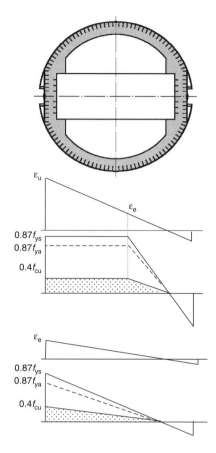

Figure 10.13 Typical stress and strain distribution across tower, at ultimate and serviceability limit states.

the internal box plates, the connector spacing is likely to be governed by the nominal spacing requirements to ensure that local buckling of the thin plates cannot occur. For this structure the connectors are at $15t_f$ (300 mm) spacing, and 300 mm long connectors are used to ensure a firm connection to the concrete.

Prestressed steel–concrete composite bridges

...for the composite section with corrugated webs...the reduction in dead load and section area will lead to a reduction in prestress...

Introduction

The prestressing of a structure involves the application of a state of stress to improve its behaviour. The prestress may be by applying displacements to the supports or by applying load from pre-strained or tensioned steel strand. Pre-stressing using high-tensile steel strand or bars is common for concrete structures, the use of strand with an ultimate strength of $1880\,\mathrm{N/mm^2}$ (Table 1.2) allows a significant reduction in steel volume when compared with conventional bar reinforcement with its $460\,\mathrm{N/mm^2}$ yield strength. This fourfold reduction allows sections to be smaller and lighter with less congested detailing. The prestressing of steel–concrete composite structures is less common [72]; however, it can be seen that there is scope for reducing the size of the conventional $355\,\mathrm{N/mm^2}$ steel plate. The main disadvantage of prestressed steel–concrete composites is that tendon anchorages or deviators attached to the steel elements tend to significantly increase fabrication complexity and so increase cost.

Displacement of supports

The raising or lowering of supports may be used to change the stress in a continuous structure. For a two-span continuous girder of constant section, move-ment of the central support vertically (δ) will induce a moment of:

$$M = \frac{3E_a I_{a-c}}{L^2}\delta \qquad (11.1)$$

For a multi-span structure the use of support movement to change moments is more complex. To induce a uniform moment along the beam each support would be moved by a different amount, initially the beam would be constructed on jacks such that a constant radius of curvature (r_j) was formed, the section would then be lowered to the final curvature (r_f):

$$M = \frac{E_a I_{a-c}}{(r_f - r_j)} \qquad (11.2)$$

For both Equations (11.1) and (11.2) the girder stiffness ($E_a I_{a-c}$) is a critical parameter. For a steel–concrete composite structure the girder stiffness will depend upon the age and state of stress of the concrete element. The moment initially jacked into the section will reduce as creep changes the effective stiffness of the section (Equation (1.3)). For use in prestressed sections the effective elastic modulus is estimated using a modified form of Equation (1.3):

$$E'_c = \frac{E_c}{(1 + k\phi)} \tag{11.3}$$

where ϕ is the creep factor (Chapter 1) and k is a modification coefficient that depends on the variation of stress with time and the properties of the section. For prestress by jacking $k = 1.5$, for prestressing using tendons $k = 1.1$.

Prestress using tendons

Prestressing of a structure using tendons will usually be one of two types, bonded or unbonded. For bonded prestress the tendon is enclosed within and bonded to the structure. For unbonded prestress the tendon is unbonded from the structural components, only touching the structure at anchorages and intermediate deviators (this unbonded, sometimes called external prestress, may be inside a box girder to improve durability).

Bonded prestress is usually used in the concrete element of a structure, as it is difficult to get adequate bond between prestressing tendons and a steel structure. For bonded prestress the concrete is cast with ducts (empty flexible tubes) inside it to the required profile, with an anchorage casting on each end. The strands are threaded through the ducts when the concrete has cured. The strand is stressed from one of the anchorages (the live end) while the other end is held at the other anchorage (the dead end); once the required force has been stressed into the strand the anchorage is locked off. The duct is then grouted to provide protection to the strand. For durability the grouting should be carried out carefully and the material used have a low bleed (separation of water from the grout mix) [134]. This method has been used (mainly in Europe) to provide prestress over supports, to induce compression in the slab and avoid cracking.

Unbonded prestress again utilises tendons, the loads are only applied to the structure at anchorage and deviator locations. The anchorages may be embedded in the concrete (similar to anchorages for bonded prestress) or attached to the steel element by a series of plates and stiffeners. Deviators are located at intermediate locations along the structure, usually at not more than 12 m centres to avoid possible vibration of the tendons. The tendons change direction at the deviators, and the tendons must change direction in a gentle curve, the radius of curvature depending on the tendon size. The deviators are again fabricated plates and stiffeners on the steel elements, or reinforced concrete blocks on concrete elements. Steel or polyethylene tubes link the anchorages and deviators. Again the tendons are threaded through the tubes and stressed from the live anchorage. For durability the tendons are grouted, the grouting medium may be a cement-based grout as the internal tendons or a petroleum wax-based medium. With the wax-based medium the tendons could be de-stressed and replaced, provided sufficient strand is left in the anchorage cap.

The primary difference between bonded and unbonded prestress is that bonded prestress is physically bonded to the structure by grout, such that at the ultimate limit state, if the concrete is strained sufficiently, the tendon stress will increase up to the ultimate strength. For unbonded prestress there is no bond and very little additional strain; at the ultimate limit state the force in the tendon will not increase far beyond that already in the tendon. An advantage of external prestress is that the tendons are accessible and can be inspected or replaced.

Design of prestressed composite structures

For a non-prestressed composite structure the stresses on the section are summed at the various stages of construction as the load is applied (Fig. 11.1(a)). For a prestressed composite structure additional stresses are induced, tending to increase the range of the live load that can be added (Fig. 11.1(b)). The additional prestressing stresses are formed from an axial and a bending component. The axial stress is simply the prestress force divided by the area of the composite section:

$$f_{PA} = \frac{P_o}{A_{ac}} \tag{11.4}$$

The bending stress is derived from the moment applied by the prestress, the moment being the prestress force multiplied by the eccentricity of the force from the section neutral axis (although more complex for continuous structures [140]):

$$f_{PZ} = \frac{P_o e}{Z_{ac}} \tag{11.5}$$

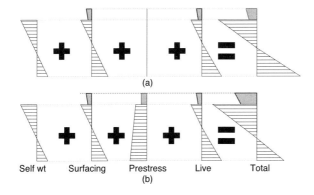

(a)

Self wt   Surfacing   Prestress   Live   Total
(b)

Figure 11.1 Comparison of stresses in a conventional composite section and a prestressed composite section.

Table 11.1 *Limiting stresses for prestressed steel–concrete composite structures*

Stage	Ultimate limit state	Serviceability limit state
During concreting	$f_a < 0.95 f_{ac}$ $f_a < 0.95 f_y$	$f_a < f_y$
At transfer	$f_a < 0.95 f_{ac}$ $f_a < 0.95 f_y$ $f_c < 0.5 f_{cu}$ $f_c > 0.5 f_{ct}$ $f_p < 0.95 f_{pu}$	$f_a < f_y$ $f_c < 0.4 f_{cu}$ $f_c > -1\,\text{N/mm}^2$ $f_p < 0.8 f_{pu}$
With full loads	$f_a < 0.95 f_{ac}$ $f_a < 0.95 f_y$ $f_c < 0.4 f_{cu}$ $f_p < 0.95 f_{pu}$	$f_a < f_y$ $f_c < 0.3 f_{cu}$ $f_c > -2.5\,\text{N/mm}^2$ $f_p < 0.7 f_{pu}$

Like other steel–concrete composites consideration of each stage of construction needs to be made. For prestressed composites the stage where prestress is transferred to the structure may be critical. Both ultimate and serviceability limit states again need consideration; Table 11.1 outlines the limits of stresses in a prestressed girder. The shear interface will also require checking at the serviceability limit state with the additional longitudinal shear induced by the prestress. It may be assumed that this is transferred over a length L_s in a similar way to that outlined for a cable-stayed anchorage (Chapter 10) using Equation (3.13).

Prestress losses

An initial force (P_i) is stressed into the tendons during construction and transferred to the bridge. Some loss of prestress will occur straight away, more will occur over time to leave a lower force (P_o). The primary causes of loss are slip, relaxation, elastic shortening, wobble, cable curvature, shrinkage and creep. The loss of prestress (α) is expressed as a percentage of the initial force:

$$\alpha = \frac{100}{P_i}(P_i - P_o) \tag{11.6}$$

The anchorage of tendons typically is via a system of conical wedges that grip the individual strands [140]. When initially installed these wedges may slip by a small amount (Δ), usually 4 to 6 mm. The loss is related to the strand area (A_s), elastic modulus (E_s) and length (L_{st}) and slip:

$$\alpha_{sl} = \frac{100 E_s A_s}{P_i L_{st}} \Delta \tag{11.7}$$

Relaxation of the strand will occur, the amount being dependent on the type of strand and amount of load applied. The losses are approximately 10% for a standard strand and 3% for a low-relaxation strand when stressed to 75% of the ultimate tensile strength of the strand.

As the tendon is stressed the composite girder will shorten slightly, causing a loss of prestress, related to the relative area of the tendon and the short-term composite section (A_{acs}):

$$\alpha_A = \frac{95 A_s}{A_{acs}} \tag{11.8}$$

For bonded tendons there may be some misalignment or movement of the duct during concreting; this misalignment or wobble will cause some loss due to friction (μ). For unbonded tendons wobble may be ignored unless the deviator length is a significant proportion of the cable length:

$$\alpha_w = \sum 100 \mu x_s \tag{11.9}$$

As the tendon changes angle (θ) at deviators or within the duct a further loss will occur due to friction with the duct at any distance x_s from the jacking end:

$$\alpha_\theta = \sum 100(1 - e^{-\mu\theta})$$

Shrinkage of the concrete will cause a shortening of the composite section (Chapter 4), using a shrinkage strain of 200×10^{-6} the loss of prestress is:

$$\alpha_{sh} = \frac{A_s E_s A_{ac}}{50 P_i A_c} \tag{11.10}$$

Over time, creep of the concrete will occur, shortening the section and shedding load from concrete to steel and causing a loss of prestress related to the relative area of the tendon and the long-term composite section (A_{acl}):

$$\alpha_{\text{cr}} = \frac{95 A_{\text{s}}}{A_{\text{acl}}} - \alpha_{\text{E}} \qquad (11.11)$$

Example 11.1

Ah Kai Sha Bridge [89, 128] is a 704 m-long, cable-stayed, double-deck, truss bridge with a 360 m main span. The upper deck is 42 m wide and carries a dual four-lane superhighway, the lower deck carries six lanes of local roads. Transversely the deck consists of a prestressed steel–concrete frame structure spanning from the outer stay locations and supporting the longitudinal trusses (Fig. 11.2). The primary prestressed steel–concrete frames are at 7 m centres and are in line with the verticals of the longitudinal truss. A series of steel–concrete composite stringer beams span between the frames.

Loading and analysis

The bridge was designed to Chinese standards with verification to UK standards. The dead load of the structure is approximately 6.5 MN per frame and the characteristic live load 2.0 MN. Analysis of the structure was by a series of longitudinal and transverse frames with local finite element analysis. The detailed analysis confirmed that the main prestressed beam behaved almost as a simply supported beam. The prestressed section is considered at both the ultimate and serviceability limit state (Fig. 11.3). Three primary load cases are considered: two during construction, before prestressing and when the prestress is added but the full dead load of the upper deck has not been placed, and finally the condition with full dead and live loads.

Prestressing of the section consists of two tendons each comprising 19 strands; each strand has an ultimate tensile strength of 265 kN. The tendons are anchored at each stay location and run through the inclined struts and along the lower beam; they are deviated at the beam strut intersection and at midspan. The

Figure 11.2 Deck cross-section for Example 11.1 (© Benaim).

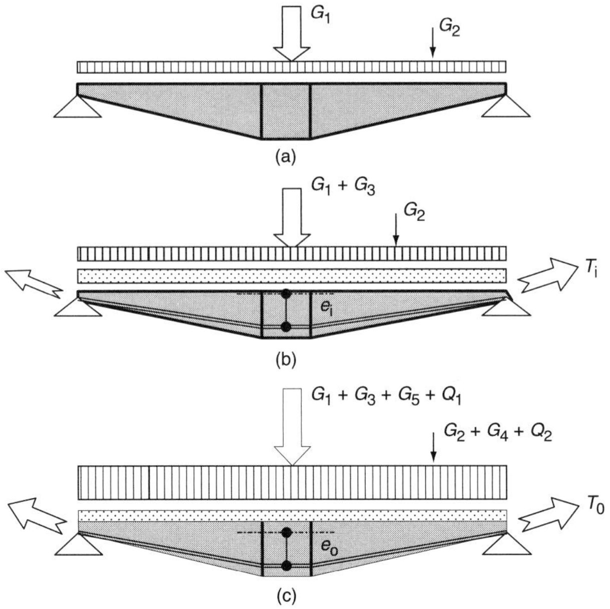

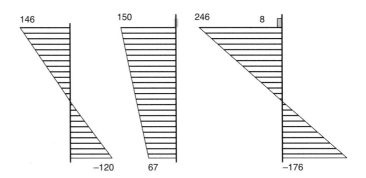

Figure 11.3 Simplified loading at various stages: (a) non-composite, no prestress, during concreting; (b) short-term composite at prestress transfer; (c) long-term composite with prestress after losses.

Table 11.2 Prestress losses for Example 11.1

Type	Amount: %	Remarks
Slip	–	Slip occurs only at ends, not beyond initial deviator
Relaxation	3	Low relaxation strand
Wobble	–	None for unbonded tendons
Curvature	15	Friction coefficient, 0.05 to 0.2
Shrinkage	1	
Creep	2	
Total	21	$T_o = 0.79 T_i$

tendons are initially stressed to 75% of the ultimate strength giving an initial prestressing force of 7.6 MN. Loss of prestress will occur, for this example the losses are summarised in Table 11.2, the total losses being 21%. The girder is composed of a fish-belly girder with the deepest section at midspan. The stresses at the girder midspan for the three primary stages are summarised in Fig. 11.4.

Figure 11.4 Stresses at various stages for Example 11.1.

Durability

For all structures durability is an issue, and for prestressed structures with steel at permanently high stresses particular care should be taken to ensure durability of the strand and its anchorage system. For prestressed structures a multi-layer philosophy is used [134], more layers of protection are added for structures at higher risk, in more severe environments.

The elements of a multi-layer protection are numerous. The nature of the bridge, the use of continuity and avoidance of joints will improve durability, thus keeping the tendons away from direct contact with chloride-contaminated water. The use of a low-permeability concrete with low crack widths will be of benefit as will a high-quality, sprayed waterproofing material. Placing the tendons within a box section will lower the exposure category; however, care must be taken at the end anchorage, as these are likely to be in a more severe environment. At an anchorage near a joint the anchor head should be kept as far away from the joint as possible and an additional coat of paint (on steel) or waterproofing layer (to concrete) should be provided to the anchorage area. Over the length of the tendon the strand must be enclosed in a sealed non-corroding duct. Within the duct and any void behind the anchorage cap a grout or wax protection should be used.

Prestressed composite box girders

The use of prestressed concrete box girders is common; they are often very similar in terms of cost of construction when compared with a conventional steel–concrete composite girder and slab construction. Prestressed composite box girder bridges combine both forms of construction. Initial designs for prestressed composite girders used concrete flanges with steel plate webs, the prestress being external but within the box. For this form of structure the steel web is proportionally stiffer than the concrete and attracts a significant proportion of the prestress from the flanges, lowering the structural efficiency. A more recent development is the use of folded plate webs, which have a low axial stiffness and attract almost no prestress. The folded plate web also has a good transverse stiffness to resist distortion of the box (see Chapter 7).

For the composite prestressed box with a folded plate web, the top and bottom flanges resist the bending effects, the webs carry all the shear, and there is little interaction between the two. A shear connection will be required between the web and flanges. Using Equation (1.16) and assuming that the second moment of area (I_{a-c}) is:

$$I_{a-c} = \frac{2A_c y^2}{b} \tag{11.12}$$

$$Q_1 = \frac{V A_c y}{I_{a-c} n} = \frac{V}{2y} = \frac{V}{d} \tag{11.13}$$

As perhaps expected, this indicates a similar shear flow longitudinally and vertically.

Corrugated webs

The corrugated steel webs carry the entire shear in a composite steel–concrete box and almost no other load [135]. The shear capacity of the web will be the lower of the yield or the buckling strength (V_{ai}) of the web. The shear capacity at yield is similar to other steel plate structures (Equation (1.9)):

$$V_y = 0.55 t d f_y \tag{11.14}$$

The buckling strength of the web is determined from an interaction of local (V_{ab}) and overall (V_{ag}) buckling of the panel:

$$\frac{1}{V_{ai}} = \frac{1}{V_{ab}} + \frac{1}{V_{ag}} \tag{11.15}$$

Local buckling of one of the folded panels may occur, and it is assumed that the geometry of the folds is such that the critical length is that of the panel parallel to the girder axis:

$$V_{al} = \frac{1.1 t^2 d (E_a f_y)^{0.5}}{b} \tag{11.16}$$

Overall buckling of the web will again depend on the geometry of the folded plate, primarily the distance across the folds:

$$V_{ag} = \frac{55}{d} (K_x K_z^3)^{0.25} \tag{11.17}$$

$$K_x = \frac{E_a t^3}{12} \left(\frac{L_1}{L_o} \right) \tag{11.18}$$

$$K_z = \frac{E_a b t s^2}{4 L_1} \tag{11.19}$$

The overall buckling limit above is for a composite box structure with relatively stiff flanges, if a steel plate flange is used the web capacity should be halved. The geometry and notation for the corrugated web are outlined in Fig. 11.5.

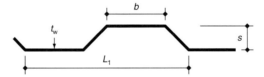

Figure 11.5 Geometry of a corrugated web plate.

Example 11.2

The A13 viaduct is a prestressed box girder structure constructed in east London [61]. For this example a comparison is made with the constructed concrete box and an alternative prestressed steel–concrete composite box using corrugated webs. The structure is continuous for 1754 m and consists of a series of spans typically 64 m in length. Each box is 14 m wide, two boxes are placed side by side to form the typically 28 m-wide structure.

For the concrete section the area of the box is 7.1 m^2 and the dead load on a span is approximately 11.5 MN. For the composite section with corrugated webs the area reduces to 4.3 m^2 and the dead load decreases by 17% to 9.5 MN. The reduction in the dead load and the area of the structural section will lead to a reduction in prestress requirements.

In this example a 5 MN ultimate shear occurs on the web near the quarter span location. A 15 mm web (t_w) with a fold thickness (s) of 350 mm and flat length (b) of

600 mm is used. From this L_1 is calculated as 1900 mm and L_o as 2187 mm. The depth of the web between flanges is 1830 mm. Using Equation (11.14) the limiting shear capacity is estimated:

$$V_y = 0.55tdf_y = 0.55 \times 0.015 \times 1.83 \times 355 = 5.3 \,\text{MN}$$

The local buckling capacity is determined from Equation (11.16):

$$V_{al} = \frac{1.1t^2d(E_a f_y)^{0.5}}{b} = \frac{1.1 \times 15^2 \times 1830(210\,000 \times 355)^{0.5}}{600} \times 10^{-6} = 6.5 \,\text{MN}$$

The global buckling capacity is derived from Equations (11.17) to (11.19):

$$K_x = \frac{E_a t^3}{13}\left(\frac{L_1}{L_o}\right) = \frac{210\,000 \times 15^3 \times 1900}{12 \times 2187} = 51.3 \times 10^6 \,\text{N mm}$$

$$K_z = \frac{E_a bts^2}{4L_1} = \frac{210\,000 \times 600 \times 15 \times 350^2}{4 \times 1900} = 30.5 \times 10^9 \,\text{N mm}$$

$$V_{ag} = \frac{55}{d}(K_x K_z^3)^{0.25} = \frac{55}{1.83}(51.3 \times 10^{-3} \times 30.5^3)^{0.25} = 185 \,\text{MN}$$

By inspection the buckling interaction will be small and the yield capacity of the web is the governing criteria. The 15 mm web is adequate at the quarter point but a thicker section will be required nearer the support.

Extra-dosed bridges

This form of structure is a hybrid between cable-stayed and externally prestressed bridges. The structure consists of a stiff deck often constructed in cantilever using a composite steel–concrete with corrugated webs and low external cables. For a cable-stayed bridge the deck is relatively flexible and a large proportion of the load is carried by the cable stays, often resulting in a large stress range, leading

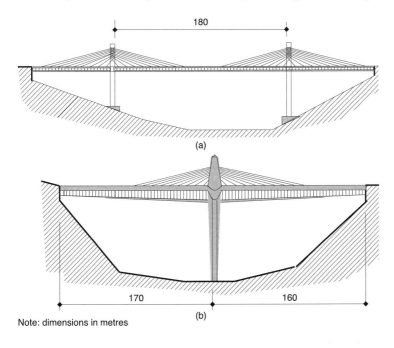

Note: dimensions in metres

Figure 11.6 Japanese extra-dosed bridges: (a) Himi Bridge with 180 m main span; (b) Rittoh Bridge with 160 m spans.

to fatigue governing the stay. For prestressed bridges the strand forces are higher but there is little variation in the tendon force due to live loads. An extra-dosed bridge has some prestress cables arranged externally above the deck, like a cable-stayed system but with much lower towers. The arrangement must be carefully chosen to ensure the girder stiffness dominates that of the cables, such that it carries all of the live load. A number of extra-dosed composite bridges with corrugated webs have been constructed, particularly in Japan [129] (see Fig. 11.6).

Assessment of composite bridges

...sometimes we underestimate the full range of ways a structure can behave...

Introduction

The approach to assessment is different to that of design [132]. In design, the engineer has the freedom to choose a form and provides steel, concrete and connecting elements to suit the forces developed from an analysis of the applied loads. Where areas are found to be critical the design can be enhanced. For assessment, the assessing engineer has no choice in the size of elements or the distribution of materials through the structure, another designer took that decision long ago. The assessing engineer has to respect the decisions taken and look for the load paths used. Often the structural types are different from those that are economic nowadays, many are simple assemblages of beams and concrete not readily analysed by modern design tools. Sometimes we underestimate the full range of ways a structure can behave, the assessor should take care not to condemn a structure simply because of a lack of understanding of its real behaviour. We tend to think we know things with a greater certainty than we really do. An example is the behaviour of deck slabs under wheel loads, the analysis using plate bending theory often shows existing slabs are underreinforced, even yield line analysis may be conservative. Research has shown that in-plane arching action may dominate shorter spans and that even larger slab panels will have some degree of arching [68]; the use of arching usually shows that a much higher load can be carried by the given reinforcement [46].

History

Iron has been used in building bridges since 1779, with the building of iron arch over the River Severn at Coalbrookdale [130], and it was used in the first long-span box girder over the Menai in 1850 [70]. The first use of steel on a major bridge was Eads triple arch over the Mississippi at St Louis in 1874 [108]. The first major use of steel in the UK was the cantilever across the Forth estuary in 1890 [109]. None of these bridges used concrete in the superstructure, the decking was typically timber baulks.

Concrete has been used in bridges for about two thousand years, concrete and masonry was used to form many Roman arch structures [97]. The modern development of concrete started in the 1800s, initially utilising arch structures for relatively modest spans. In the 1850s, with the patenting of reinforced concrete by Wilkinson, there was an era of experimentation with the addition of various bar arrangements or by utilising embedded iron or steel sections. The development of reinforced concrete in the UK in a recognisably modern form was by Hennebique and Mouchel in the early 1900s, and many bridges at that time had an arch form (of open spandrel construction).

Iron joist sections embedded in concrete using a filler joist or jack arch form were used in fireproof slabs from the 1850s. The first recorded bridge with a combination of iron and concrete in the UK was the River Waveney Bridge [133], a 15 m-span structure comprising a wrought iron arch with a braced spandrel, infilled with mass concrete, dating to 1870. With the reconstruction work at Paddington in 2003, an iron-rib and concrete arch by Brunel dating to 1838 was rediscovered [138]. The unusual shaping of this structure leads to speculation that this shape was conceived to utilise some interaction between the iron and concrete. Many of these innovative bridges utilise an arch form. It may be postulated that this was to reduce risk. The way an arch works can be estimated using simple geometry, it is well tried and tested, the use of the new materials will then be the only real unknown.

The use of embedded-joist sections (Fig. 12.1(a)) in concrete is probably the first recognisably modern use of composite steel–concrete construction. The development of steel and concrete remained largely separate for the first half of the twentieth century, steel bridges being generally preferred for railway bridges and longer-span road structures, concrete being used on shorter spans, primarily carrying roads. When steel was combined with a concrete deck a composite section was not utilised, usually the steel element was designed to carry all loads. The concrete element acting as a deck slab only carried local wheel loads. The

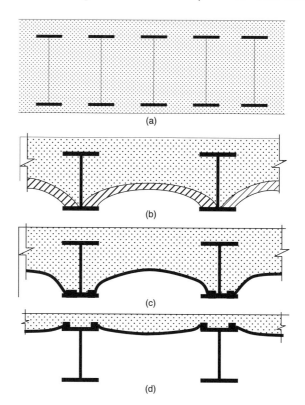

Figure 12.1 (a) Embedded joists; (b) brick jack arch; (c) steel hog plates; and (d) buckle plate deck.

use of longitudinal stringers supporting the deck above the main steel ensured a limited interaction. Where the steel and concrete were in contact, the provision of composite action was often only accidental, for instance from the joints and connections in buckle plates (Fig. 12.1(b)–(d)). Where composite action between steel and concrete was assumed the steel element was usually fully encased (see Figs 9.2 and 12.5(b)).

Structure types

The structural forms from simple beams and boxes, through to arches and suspended spans outlined in Chapter 1 have all been used before and may need to be evaluated in an assessment. The details of these structures may be different to those contemplated nowadays. Many girders will not be formed from rolled sections or welded plate but fabricated from plates and angles using rivets, often using multiple thin plates to achieve the required thickness. Trusses and arches are not simple frameworks of large steel sections or filled tubes but lattices of battened members with a myriad of intermediate braces in all directions, often forming a very complex structure.

Slabs may be formed from brick arches filled with concrete or use steel plates (Fig. 12.1(b) and (c)), clearly showing the assumption of arching rather than bending in the original design. Other slabs may use buckle plates or filler beams. On many larger structures, the concrete slab may not be fully effective due to inadequate shear connection (by modern standards). The slab may be on a series of stringers separated from the primary structure and only partially effective as a composite element.

Inspection

Viewing a structure will give an indication of its overall integrity and condition. A detailed inspection of the individual components and measurement of dimensions will enable accurate estimates of loading and any loss of section. If possible the bridge should be viewed when fully loaded such that deflections, vibration or the rattling of loose elements can be observed. Inspection just after a period of heavy rain can also be useful, allowing observation of leaking joints and cracks.

The steel elements should be inspected for defects including corrosion, cracking, loose or missing bolts and signs of buckling or out-of-flatness in slender plate panels. For rail bridges, in particular, details that are known to be fatigue susceptible should be more carefully examined for cracking.

The concrete elements should be inspected and any cracking, rust staining, spalling or crazing noted. Where chloride contaminated water penetration is evident, consideration of testing such as a cover survey to reinforcement, concrete sampling to determine the extent of carbonation and chloride content, and the use of a half cell and resistivity measurements [139] to determine any active corrosion pockets may also be useful. At the interface of the steel–concrete composite elements signs of corrosion, separation, cracking or longitudinal slip should be noted.

Loads

The loading used for assessing a structure is likely initially to be the full standard highway or railway loads (Chapters 2, 4 and 10). If the structure is not capable of

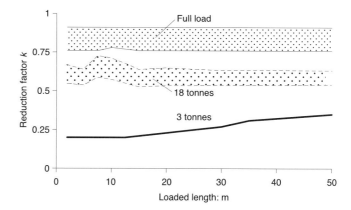

carrying this load then the load that it can carry should be determined. The determination of the load limit is carried out on a structure-by-structure basis, as the loading will be influenced by a number of bridge-specific factors.

Highway loading

An HA-type loading approximates 40 tonnes vehicle loading, and it is derived using worst credible values of vehicle overloading and bunching; for situations where this is not likely some reduction in loading intensity may be used. Current regulations [19] define three categories of structure based on traffic flow (high, medium and low), each of these is further subdivided depending on the condition of the surfacing (good or poor). Figure 12.2 shows the variation of the reduction factor (k) from standard loading. The reduction ranges from 0.91 for high traffic flows with poor surfacing to 0.76 for low flows and good surfacing.

If the bridge is still unable to carry this modified full loading then a reduced loading is considered, usually 26, 18, $7\frac{1}{2}$ or 3 tonnes loading. Figure 12.2 also shows the reduction factors for these loads, the 3 tonnes limit giving a reduction factor of up to 0.2. As an alternative to weight restrictions it may be possible to reduce carriageway widths to limit the number of lanes of full loading on the structure.

Railway loading

For railway bridges in the UK the assessment is normally carried out for 20 units of RA1 loading (Fig. 12.3(a)) [131]. For bridges where the trains pass at more than 8 kilometres per hour (kph) the load should be increased by a dynamic factor (k_D):

$$k_D = 1 + \phi \tag{12.1}$$

where ϕ is a dynamic increment depending on the train speed, train–structure interaction, structure length and the natural frequency of the bridge. Figure 12.3(b) outlines the range of the dynamic increment for various spans and train speeds. The number of units of RA1 loading that can be carried by the structure is determined:

$$N^o_{RA1} = \text{effects of 20 units of RA1} \times k_D k_C \tag{12.2}$$

Typically at least two capacities should be used: first, assuming the load reduction factor (k_C) is 1.0, the limiting value of k_D for 20 units can then be determined;

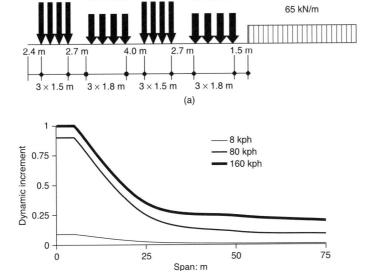

*Figure 12.3 (a) Type
RA1 loading (20 units);
(b) range of dynamic
increment for railway
assessment loading with span
and line speed.*

second, the dynamic factor for the defined line speed is used allowing a value of k_C to be determined. These calculations will result in an estimate of the speed restrictions to be placed on heavier freight trains (say 90 kph for 20 units) or an estimate of the weight of lighter passenger vehicles that can travel at the full line speed (say 15 units at 160 kph).

Materials

The behaviour of the composite structure is heavily influenced by the properties of the component materials. For older structures the properties of the iron, steel and concrete may be different to those assumed nowadays. Strengths are usually less, the variability in strength is often an issue, and quality control procedures on some older structures were poor.

Concrete

For pre-1939 bridges a concrete strength of $12/15 \, \text{N/mm}^2$ may be assumed, for younger structures $21/25 \, \text{N/mm}^2$ strength is appropriate, unless testing is carried out to determine the strength more accurately. The lower strength of the concrete will affect the elastic modulus, a long-term value (E_c') of $14 \, \text{kN/mm}^2$ is recommended [95]. The shear strength (v_c) using Equation (5.9) is reduced by a factor of:

$$\left(\frac{f_{cu}}{40}\right)^{1/3} \tag{12.3}$$

The capacity of any shear connection will be reduced, using Equations (1.19a), (1.20) and (1.21) it can be seen that this will be in direct proportion to the lower strength.

Steel

For older structures there may be a considerable variation in strength. For the Eads bridge of 1870, the ultimate strength varied from 400 to 900 N/mm^2 [108], typically a 230 N/mm^2 yield strength may be assumed. For more modern structures, yield strengths above 250 N/mm^2 will be dependent on the grade of steel used and unless clearly shown on record drawings or determined from tests a conservative value is initially assumed.

Testing of the shear connection

The strengths of the standard connectors outlined in Table 1.5 and Equations (1.19) to (1.23) have been calibrated by tests, the newer forms of perforated plate connectors are also tested as part of the development process [26]. The standard test [4] comprises a push-out test on a pair of connectors (Fig. 12.4(a)). At least three tests are required and the lowest result is taken as the nominal static strength; where more tests are carried out a statistical estimation of the static strength can be calculated. The surface of the steel beam in contact with the concrete is normally greased to ensure that the bond between steel and concrete does not influence the test results. The standard test can underestimate actual connector strengths, particularly for connectors with limited resistance to separation. A modified and larger test [9, 95] is used for these situations (Fig. 12.4(b)). The test will more accurately reflect the structure geometry and connector capacities.

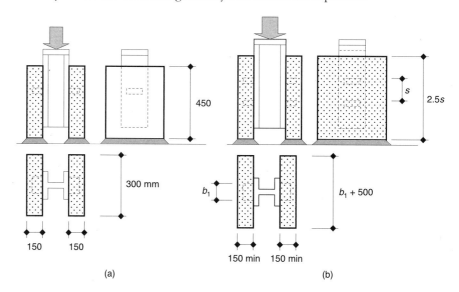

*Figure 12.4 Test layout for
connectors: (a) standard
test; (b) special test.*

Analysis

The analysis of an existing structure being assessed will use similar methods to those used in design [45]. For most structures an elastic analysis method is acceptable; however, where the composite section is not slender and is capable of some plastic deformation prior to failure then some redistribution of moments may be assumed to occur. For steel elements being assessed it is usually acceptable to use the section modulus calculated at the centre of the flange rather than the extreme edge.

Incidental and partial composite action

Many older bridges may not have full connection complying with current design requirements. Connectors, where provided, may not be of modern types, with connectors spaced uniformly along the girder rather than increasing towards the supports. The connectors may also be grouped adjacent to stiffeners or cross-beams with little connection between them. Often part of the shear connection may be only incidental.

Incidental shear connectors are ends of plate, bolt or rivet heads, stiffeners or trough connections, none of which were specifically designed as connectors but nevertheless provide some connection. Incidental shear connection should not usually account for more than 25% of the longitudinal shear capacity [95]. Incidental shear connection usually has low resistance to uplift. An estimate of the connector capacity can be made by using Equations (1.19a) to (1.23) as appropriate, whichever form is nearest.

For girders of less than 20 m span, if insufficient connection is provided to give complete connection then a reduced strength may be derived assuming partial connection [9]. Partial interaction is usually expressed in terms of the maximum design load that the fully composite span could carry:

$$W_D = G_D + Q_D \tag{12.4}$$

For ductile or flexible connectors such as studs or channels, where slip can occur and redistribute load along the connection, then the capacity is determined by assuming a linear increase in capacity from that of the steel element alone (W_a) to the full design capacity (Fig. 12.5). The reduced design load (W) is determined by interpolation from the proportion of the full connector requirements. For stiff connectors where less slip occurs the capacity of the section can again be assumed to increase linearly, but from a lower base, either the design dead load (G_D) for unpropped construction or from zero load for propped construction (again shown in Fig. 12.5).

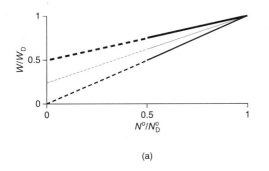

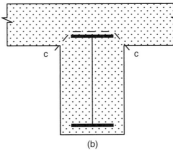

Figure 12.5 (a) Partial shear connection, typical variation in span load (W) with the number of connectors (No); (b) cased beam showing shear plane used to assess the capacity.

Cased beams

Fully encasing the steel section with concrete was a popular method of forming early composite structures, generally shear connectors were not used and the composite action relies on bond between the steel and concrete. When assessing such structures the capacity of the section will often be limited by this bond stress, particularly those sections that are compact and might otherwise develop the full plastic capacity of the section. For cased beams, a check of the longitudinal shear across the top of the steel section is carried out (Fig. 12.5(b)). For design, the bond stress at this interface would normally be limited to $0.4\,\text{N/mm}^2$ at the serviceability limit state. For assessment a less conservative limit is used, the lesser of Equations (12.5a) and (12.5b):

$$Q_1 = 0.1(f_{cu})^{0.5} L_s \qquad\qquad\qquad (12.5a)$$

$$Q_1 = 0.7 L_s \qquad\qquad\qquad\qquad\quad (12.5b)$$

If the cover around the steel section is less than 50 mm the limiting strength of the interface should be reduced, by 30% for covers of 20 mm.

At the ultimate limit state where no reinforcement crosses the shear plane the capacity of the shear plane is limited to:

$$Q_1 = 0.5 L_s \qquad\qquad\qquad\qquad\quad (12.6)$$

Where reinforcement passes through the shear plane the capacity may be assessed using Equation (1.24):

$$Q_1 = 0.9 L_s + 0.7 A_s f_y \qquad\qquad\quad (1.24)$$

The minimum amount of reinforcement passing through the shear plane should be:

$$A_s = \frac{0.8 L_s}{f_y} \qquad\qquad\qquad\qquad\quad (12.7)$$

Strengthening

The primary cause of deterioration in bridges is from chloride-contaminated water ingress at joints, the replacement or removal of joints being the best long-term repair option. Where corrosion has affected the structure's capacity and it can no longer carry the required loading then strengthening and upgrading will be required. The strengthening of concrete elements can be achieved through the bonding of additional steel or carbon fibre reinforcement to it with epoxy [139]; in some circumstances the addition of an over-slab bonded to the original slab may be applicable. For steel elements, strengthening is often relatively simply achieved by the addition of extra plates and stiffeners using welding or bolting techniques as applicable. For both steel and concrete elements, prestressing by jacking of supports or the use of external unbonded tendons has been successfully carried out [58]. Where the interface connection is weak and requires strengthening repair options are more limited. In some situations it may be possible to break out limited areas of concrete and attach connectors to the steel element in groups. Where there is also concern over the integrity of the deck slab due to chloride ingress or carbonation of the concrete then the complete removal of the deck may be the only solution, short of complete demolition.

Life-cycle considerations

The life cycle of a bridge starts at its initial construction, through its useful life; considering inspections, testing, repair or strengthening, up to its final demolition. Decisions made during the initial design will influence significantly the behaviour through the life cycle. A whole-life cost [71] of the bridge through its life cycle may be obtained by placing values on the various operations throughout the structure's life; these may use discounted cost methods (Chapter 4, Equation (4.1)). More recently the environmental life cycle has been considered [77], looking at the embodied energy and carbon dioxide (CO_2) emissions through the cycle. Both types of life-cycle studies indicate that costs caused by disruption to the use of the structure outweigh the initial construction cost by a significant margin, this is particularly true for motorway or main line railway bridges but is also applicable to smaller structures where the length of any diversion around the structure is significant. Consequently when planning inspections, testing, maintenance, repair or the rebuilding of a bridge, the minimum disruption to its use should be made.

Appendix A: Approximate methods

For a girder with approximately equal flange areas (A_f), overall depth D, and web thickness t_w, the primary second moment of area of the section may be approximated as:

$$I_a = 2A_f(0.5D)^2 + \frac{D^3 t_w}{12}$$

(A1a)

or rearranging Equation (A1a) [34]:

$$I_a = 2\left(A_f + \frac{D t_w}{6}\right)(0.5D)^2$$

(A1b)

$$Z = \frac{I_a}{0.5D}$$

(A2)

Substituting Equation (A2) into Equation (A1b):

$$Z = (A_f + 0.167 D t_w)D$$

(A3)

$$M = Z f_a$$

(A4)

$$M = (A_f + 0.167 D t_w)D f_a$$

(A5)

So the moment that the beam will carry may be estimated from the flange area plus one sixth of the web, multiplied by the permissible stress and the beam depth. This approximation can also be used for girders of unequal flange area.

The transverse second moment of area of the section may be approximated as:

$$I_y = \frac{2B^2 A_f}{12}$$

(A6)

If the web is of a similar area to the flange:

$$A = 3A_f$$

(A7)

$$r_y = \left(\frac{I_y}{A}\right)^{0.5}$$

(A8)

$$r_y = 0.23B$$

(A9)

This approximation can be used for unequal flanges if the average width B' is used.

Appendix B:
Calculation of section properties

Section properties for steel sections

For a steel beam fabricated from a series of plates (see Fig. B1(a)) the location of the neutral axis of the fabricated section can be calculated:

$$y_b = \frac{(A_1 y_1 + A_2 y_2 + A_3 y_3)}{(A_1 + A_2 + A_3)} \qquad \text{(B1)}$$

The second moment of area of the steel section is calculated about the neutral axis:

$$I_a = I_1 + I_2 + I_3 + A_1(y_b - y_1)^2 + A_2(y_b - y_2)^2 + A_3(y_b - y_3)^2 \qquad \text{(B2)}$$

The section modulus is derived from the second moment of area:

$$Z_b = \frac{I_a}{y_b} \qquad \text{(B3)}$$

$$Z_t = \frac{I_a}{y_t} \qquad \text{(B4)}$$

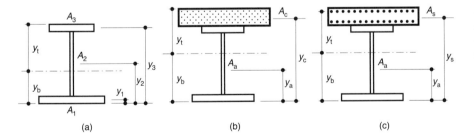

Figure B1 (a) Layout of steel section; (b) layout of composite section; (c) layout of cracked section.

Section properties for steel–concrete composite sections

For a steel–concrete composite section formed from a steel beam and concrete slab (Fig. B1(b)) the location of the neutral axis of the composite section can be calculated:

$$y_b = \frac{(A_a y_a + n A_c y_c)}{(A_a + n A_c)} \qquad \text{(B5)}$$

The second moment of area of the composite section is calculated about the neutral axis:

$$I_{a-c} = I_a + nI_c + nI_i + A_a(y_b - y_a)^2 + A_c(y_b - y_c)^2 \tag{B6}$$

The section modulus is derived from the second moment of area:

$$Z_b = \frac{I_{a-c}}{y_b} \tag{B7}$$

$$Z_t = \frac{I_{a-c}}{y_t} \tag{B8}$$

The longitudinal shear flow Q_L is calculated using Equation (1.16) and rearranging:

$$\frac{Q_L}{V} = \frac{A_c y_c}{I_{a-c}} \tag{B9}$$

Section properties for cracked steel–concrete composite sections with reinforcement

For a steel–concrete section formed from a steel beam and concrete slab where the slab is cracked (Fig. B1(c)) the location of the neutral axis of the composite section can be calculated:

$$y_b = \frac{(A_a y_a + A_s y_s)}{(A_a + A_s)} \tag{B10}$$

The second moment of area of the cracked section is calculated about the neutral axis:

$$I_{a-s} = I_a + I_s + I_i + A_a(y_b - y_a)^2 + A_s(y_b - y_s)^2 + A_i(y_b - y_i)^2 \tag{B11}$$

The section modulus is derived from the second moment of area:

$$Z_b = \frac{I_{a-s}}{y_b} \tag{B12}$$

$$Z_t = \frac{I_{a-s}}{y_t} \tag{B13}$$

For the sections used in the examples in Chapters 2 to 12 the section properties are summarised in Table C1.

Appendix C: Section properties for examples

For the sections used in the examples in Chapters 2 to 12 the section properties are summarised in Table C1.

Table C1 Section properties for examples in Chapters 2 to 12

Example	Section	A_a: m^2	I_a: m^4	y: m	Z_t: m^3	Z_b: m^3	r_y: m	n	A_c or A_s: m^2	I_{ac}: m^4	y: m	Z_t: m^3	Z_b: m^3	Z_c or Z_s: m^3	Q_1/V
2.1	Quakers Yard	0.0482	0.022	0.640	0.022	0.034	0.159	12	0.64	0.053	1.224	0.004	0.044	1.223	0.53
3.1	Blythe Bridge	0.0789	0.043	0.765	0.041	0.057	0.181	6	0.78	0.106	1.521	0.353	0.070	1.494	0.52
3.1	Blythe Bridge	0.0789	0.043	0.765	0.041	0.057	0.181	12	0.78	0.088	1.321	0.176	0.067	1.692	0.46
3.3	Nanny Bridge	0.0773	0.067	0.930	0.048	0.074	0.121	12	0.76	0.158	1.61	0.013	0.100	2.24	0.34
4.1	Walsall (mid)	0.0661	0.018	0.382	0.019	0.051	0.211	6	0.7	0.069	1.081	0.253	0.066	1.05	0.66
4.1	Walsall (mid)	0.0661	0.018	0.382	0.019	0.051	0.211	12	0.7	0.056	0.895	0.122	0.064	1.15	0.61
4.1	Walsall (pier)	0.0872	0.029	0.537	0.035	0.056	0.196	1	0.0084	0.034	0.610	0.015	0.058	0.045	
4.2	Metro (mid)	0.1453	0.295	1.804	0.177	0.165	0.173	12	0.94	0.357	1.418	0.174	0.255	5.964	0.16
								1	0.010	0.306	1.732	0.176	0.179	0.296	
4.2	Metro (pier)	0.1445	0.253	1.518	0.153	0.170	0.182	12	0.35	0.27	1.381	0.150	0.199	4.75	0.07
4.2	Metro (trans)	0.040	0.003	0.242	0.008	0.013	0.163	12	0.75	0.010	0.546	0.001	0.018	0.044	1.27
5.1	Doncaster viaduct	0.0905	0.087	0.885	0.055	0.10	0.191	12	1.12	0.219	1.746	0.01	0.127	3.156	0.36
6.1	Meghna Bridge	0.2420	1.674	3.645	0.553	0.462	0.214	12	4.00	5.75	3.645	1.900	1.586	20.71	0.13
								1	0.017	4.412	2.38	0.283	1.869	1.021	
7.1	Rail box	0.1816	0.121	0.928	0.116	0.135	0.347	1	0.003	0.125	0.945	0.048	0.136	0.118	
7.3	Sprint viaduct	0.179	0.168	1.054	0.137	0.164	0.25	12	0.7	0.249	1.386	0.043	0.184	0.244	0.24
8.1	Truss chord	0.2256	0.057	0.600	0.100	0.100	0.288								
9.3	Usk (tie)	0.250	0.244	1.679	0.307	0.147	0.372	12	3.6	0.366	2.193	1.313	0.168	10.25	0.35
								1	0.067	0.280	1.855	0.452	0.153	0.427	
9.3	Usk (arch)	0.7115	0.123	0.924	0.109	0.136	0.611	12	3.3	0.216	0.994	0.204	0.221	2.20	0.20
10.1	Second Severn	0.1356	0.097	0.670	0.064	0.154	0.229	12	4.8	0.386	1.925	1.468	0.205	10.88	0.44
11.2	AKS	0.0770	0.085	1.201	0.059	0.072	0.140	14	0.65	0.143	1.839	0.012	0.078	2.198	0.30

Appendix D:
Plastic section properties
for steel–concrete
composite sections

For a steel–concrete section formed from a steel beam and concrete slab the location of the plastic neutral axis has equal effective areas above and below it. If the axis is located in the web of the girder:

$$\frac{A_c}{n} + A_3 + A_{2b} = A_{2a} + A_1 \tag{D1}$$

The plastic modulus of the composite section is calculated about the neutral axis:

$$Z_{P_{a-c}} = A_1\,y_1 + A_{2a}\,y_{2a} + A_{2b}\,y_{2b} + A_3\,y_3 + \frac{A_c}{n}\,y_c \tag{D2}$$

If the axis is located in the concrete slab:

$$\frac{A_{c'}}{n} = A_1 + A_2 + A_3 \tag{D3}$$

The plastic modulus of the composite section is calculated about the neutral axis:

$$Z_{P_{a-c}} = A_1\,y_1 + A_2\,y_2 + A_3\,y_3 + \frac{A_{c'}}{n}\,y_c \tag{D4}$$

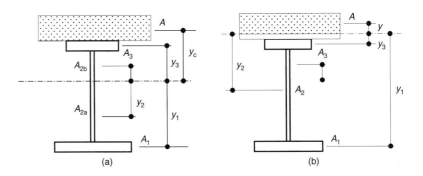

Figure D1 Layout of compact section: (a) with neutral axis in web; (b) neutral axis in concrete slab.

For the sections used in the examples in Chapters 2 to 12 the plastic section properties are summarised below:

Table D1 Plastic section properties for examples in Chapters 2 to 12

Example	Section	n	Y_p: m	Z_p: m^3	Z_p/Z
3.1	Blyth Bridge	15	1.82	0.084	1.20
4.1	Walsall Road (mid)	15	1.35	0.070	1.09
4.2	Metro (transverse)	15	0.65	0.021	1.11

Appendix E:
Torsional properties for steel–concrete composite sections

For a closed steel section made up of a series of rectangular fabrications the torsional constant J is:

$$J_a = \frac{4(BD)^2}{2Bt_f + 2Dt_w} \tag{E1}$$

For an open steel section made up of a series of rectangular fabrications that is closed by a concrete slab:

$$J_{ac} = \frac{4(BD)^2}{Bt_f + 2Dt_w + Bt_c/n} \tag{E2}$$

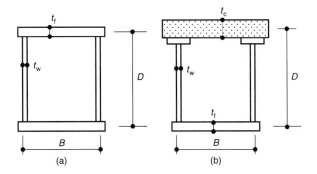

Figure E1 Layout of box: (a) closed steel box; (b) steel–concrete composite section.

For the sections used in the examples in Chapters 2 to 12 the torsional properties are summarised below:

Table E1 Torsional properties for examples in Chapters 2 to 12

Example	Section	J_a: m^2	J_{ac}: m^2
7.1	Rail box	69	—
7.3	Open box	—	203
7.3	Closed box	203	177

Appendix F: Moment–axial load interaction for compact steel–concrete composite sections

For a steel–concrete section formed from a steel section filled with concrete the moment–axial load interaction can be calculated by dividing the section into a series of elements, and considering a range of forces on the elements. The force distribution is assumed to vary from purely axial, through to a tensile load, with various intermediate arrangements involving bending (Fig. 9.6). The tensile forces on concrete elements are ignored.

For each force distribution an axial load and moment can be calculated. Plotting the results for each force distribution will enable the interaction diagram to be constructed (Fig. F1). The shape of the interaction diagram will be dependent on the contribution factor (Equation (9.3)) and the shape of the section.

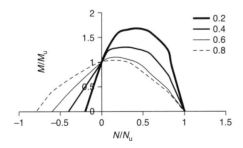

Figure F1 M–N interaction curve for a compact circular steel section filled with concrete with various steel contribution factors.

References

1. Sullivan, L. H. *The Autobiography of an Idea*, Dover Publications, New York, 1958.
2. British Standards Institution, BS 5400, *Steel, Concrete and Composite Bridges*, Part 3, Design of Steel Bridges, London, 2000.
3. British Standards Institution, BS 5400, *Steel, Concrete and Composite Bridges*, Part 4, Design of Concrete Bridges, London, 1992.
4. British Standards Institution, BS 5400, *Steel, Concrete and Composite Bridges*, Part 5, Design of Composite Bridges, London, 1978.
5. British Standards Institution, BS 5400, *Steel, Concrete and Composite Bridges*, Part 10, Design of Bridges for Fatigue, London, 1982.
6. Commission of the European Communities. *Eurocode, Common Unified Rules*. No. 1, Loads for structures.
7. Commission of the European Communities. *Eurocode, Common Unified Rules*. No. 2, Concrete structures.
8. Commission of the European Communities. *Eurocode, Common Unified Rules*. No. 3, Steel structures.
9. Commission of the European Communities. *Eurocode, Common Unified Rules*. No. 4, Composite structures.
10. Kerensky, O. A. and Dallard, N. J. The four level interchange at Almonsbury, Proceedings of the Institution of Civil Engineers, **40**, pp. 295–322, July 1968.
11. Hill, G. J. and Johnstone, S. P. Improvements of the M20 Maidstone bypass, junctions 5–8. Proceedings of the Institution of Civil Engineers, **93**, pp. 171–181, Nov 1993.
12. Narayanan, R., Bowerman, H. G., Naji, F. J., Roberts, T. M. and Helou, A. J. *Application Guidelines for Steel–Concrete Sandwich Construction: Immersed Tube Tunnels*, Technical Report 132, Steel Construction Institute.
13. Eaton, K. and Amato, A. *A Comparative Environmental Life Cycle Assessment of Modern Office Buildings*, Publication 182 Steel Construction Institute, 1998.
14. Collings, D. M5 parallel widening, *New Steel Construction*, **2**, Feb 1994.
15. Sadler, N. and Wilkins, T. Short span railway underbridges: developments, *New Steel Construction*, pp. 18–19, Nov/Dec 2003.
16. Dickson, D. M. M25 orbital road, *Poyle to M4: Alternative Steel Viaducts*, Proceedings of the Institution of Civil Engineers, **82**, Part 1, pp. 309–326.
17. Saul, R. Bridges with double composite action. *Structural Engineering International*, 1/96, pp. 32–36.
18. Lecroq, P. *Highway Bridge Decks in France*, ECCS/BCSA International Symposium on Steel Bridges, Feb 1988.
19. Highways Agency, *Design Manual for Roads and Bridges*, Volume 3, BD 21, The assessment of highway bridges, HMSO, London, Aug 2001.
20. Virlogeux, M., Foucriat, J. and Lawniki, J. *Design of the Normandie Bridge, Cable stayed and suspension bridges*, Proceedings of International Conference AIPC-FIP, Deauville, Oct 1994.
21. Hong Kong Government, Highways Department, *Structures Design Manual*, 1993.
22. AASHTO, *Standard Specifications for Highway Bridges*, 16th edition, 1996.
23. Highways Agency, *Design Manual for Roads and Bridges*, Volume 1, BD 28, Early thermal cracking of concrete, HMSO, London, March 1987.

24. BCSA, *Steelwork Design Guide to BS 5950*, Volume 1, Section properties and member capacities (4th edition), Steel Construction Institute, Publication 202.

25. Isles, D. (Ed), *Specification of Structural Steelwork for Bridges*, Steel Construction Institute, Publication 170, 1996.

26. Oguejiofor, E. C. and Hosain, M. U. Behaviour of perfobond rib shear connectors in composite beams: full size tests, *Canadian Journal of Civil Engineering*, **19**, pp. 224–235, 1992.

27. Schlaich, J., Schlaich, M. and Schmid, V. *Composite Bridges: Recent Experience. The development of teeth connectors*, Proceedings of the 3rd International Meeting, Composite bridges, Madrid, pp. 760–790, Jan 2001.

28. Highways Agency, *Design Manual for Roads and Bridges*, Volume 1, BD 10, Design of highway structures in areas of mining subsidence, HMSO, London, May 1997.

29. British Standards Institution, BS 153, *Steel Girder Bridges*, London, 1958.

30. British Standards Institution, CP 117, *Composite Construction in Structural Steel and Concrete*: Part 2: Beams for Bridges, London, 1967.

31. British Standards Institution, BS 5400, *Steel, Concrete and Composite Bridges*, Part 2, Loads for Bridges, London, 1978.

32. Highways Agency, *Design Manual for Roads and Bridges*, Volume 1, BD 37, Loads for highway bridges, HMSO, London, Aug 2001.

33. Ryall, M. *Application of the D-type Method of Analysis for Determining the Longitudinal Moments in Bridge Decks*, Proceedings of the Institution of Civil Engineers, **94**, May 1992.

34. Brown, C. *Plate Girders and Rolled Sections*, ECCS/BCSA International Symposium on Steel Bridges, Feb 1988.

35 Timoshenko, S. and Gere, J. *Theory of Elastic Stability*, Second Edition, McGraw-Hill, 1961.

36. Wang, Y. and Nethercot, D. *Ultimate Strength Analysis of Three-dimensional Braced I-beams*, Proceedings of the Institution of Civil Engineers, Part 2, **87**, Mar 1989.

37. Jeffers, E. U-frame restraint against instability of steel beams in bridges, *The Structural Engineer*, **68**, Sept 1990.

38. Rahal, K. and Harding, J. *Transversely Stiffened Girder Webs Subject to Shear Loading – part 2*: stiffener design, Proceedings of the Institution of Civil Engineers, Part 2, **89**, March 1990.

39. Hillerborg, A. *Strip Method Design Handbook*, E & F N Spon, London, 1996.

40. Highways Agency, *Design Manual for Roads and Bridges*, Volume 1, BD 57, Design for durability, HMSO, London, Aug 2001.

41. Highways Agency, *Design Manual for Roads and Bridges*, Volume 1, BA 42, The design of integral bridges, HMSO, London, May 2003.

42. Csagoly, P. F. and Lybas, J. M. Advanced design methods for concrete bridge deck slab, *Concrete International*, May 1989.

43. Highways Agency, *Design Manual for Roads and Bridges*, Volume 2, BD 33, Expansion joints for use in highway bridge decks, HMSO, London, Nov 1994.

44. England, G. and Tsang, N. *Towards the Design of Soil Loading for Integral Bridges*, Concrete Bridge Development Group, Technical paper 2, Aug 2001.

45. Hambley, E. *Bridge Deck Behaviour* (2nd edition), E & F N Spon, London, 1991.

46. Collings, D. *Design of Bridge Decks Utilising Arching Effects*, Proceedings of the Institution of Civil Engineers, Buildings and structures, **152**, Aug 2002.

47. Highways Agency, *Design Manual for Roads and Bridges*, Volume 3, BD 81, Use of compressive membrane action in bridge decks, HMSO, London, May 2002.

48. Dolling, C. and Hudson, R. *Weathering Steel Bridges*, Proceedings of the Institution of Civil Engineers, Bridge Engineering, **156**, Mar 2003.

49. Highways Agency, *Design Manual for Roads and Bridges*, Volume 2, BD 7, Weathering steel for highway structures, HMSO, London, Nov 2001.

50. Bell, B. *An Integral Composite Bridge of High Skew*, Proceedings of the Institution of Civil Engineers, Bridge Engineering, **156**, Issue BE4, Dec 2003.

51. Pucher, A. *Influence Surfaces of Elastic Plates*, Springer Verlag, 1964.

52. National Roads Authority, *Design Manual for Roads and Bridges*, Folio A: Volumes 1 and 2, Dublin, Dec 2000.

53. Isles, D. (Ed), *Replacement Steel Bridges for Motorway Widening*, Steel Construction Institute, Ascot, 1992.

54. Rockey, K. and Evans, H. (Eds), *The Design of Steel Bridges*, Granada, London, 1981.

Steel–concrete composite bridges

55. Guyon, Y. (translated by Turner, F.), *Limit State Design of Prestressed Concrete*, Applied Science Publishers, London, **2**, 1974.

56. British Standards Institution, BS 5400, *Steel, Concrete and Composite Bridges*, Part 6, Specification for Materials and Workmanship, Steel, London, 1999.

57. Walther, R., Houriet, B., Isler, W. and Moia, P. *Cable Stayed Bridges*, Thomas Telford, London, 1988.

58. Murry, M. *Friaton Bridge Strengthening*, Bridge Management 3, E&F N Spon, London, 1996.

59. Prichard, B. (Ed), *Continuous and Integral Bridges*, Proceedings of the Henderson Colloquium, Cambridge, 1993.

60. Highways Agency, *Design Manual for Roads and Bridges*, Volume 1, BD 36, Evaluation of Maintenance Costs in Comparing Alternative Designs for Highway Structures, HMSO, London, Aug 1992.

61. Collings, D. *The A13 Viaduct: construction of a large monolithic concrete bridge deck*, Proceedings of the Institution of Civil Engineers, Structures and Buildings, **146**, Feb 2001.

62. Mato, F. *Comparative Analysis of Double Composite Action Launched Solutions in High Speed Railway Viaducts*, *Composite bridges – state of the art in technology and analysis*, Proceedings of the 3rd International Meeting, Madrid, Jan 2001.

63. British Standards Institution, BS 5975, *Code of Practice for Falsework*, London, 1996.

64. Pandey, M. and Sherbourne, A. Unified v. Integrated approaches in lateral-torsional buckling of beams, *The Structural Engineer*, **67**, July 1989.

65. R. J. Roark and W. C. Young, *Formulas for Stress and Strain*, McGraw-Hill, New York, 1985.

66. Hayward, A., Saddler, N. and Tordoff, D. *Steel Bridges* (2nd edition), BSCA publication 34/02, 2002.

67. Ito, M., Fujino, Y., Miyata, T. and Narital, N. (Eds), *Cable-stayed Bridges*, *Recent Developments and Their Future*. Elsevier Science Publishers, Amsterdam, 1991.

68. Peel-Cross, J., Rankin, G., Gilbert, S. and Long, A. *Compressive Membrane Action in Composite Floor Slabs in the Cardington LBTF*, Proceedings of the Institution of Civil Engineers, Structures and Buildings, 2001.

69. Collings, D., Mizon, D. and Swift, P. *Design and Construction of the Bangladesh–UK Friendsip Bridge*, Proceedings of the Institution of Civil Engineers, Bridge Engineering, **156**, Dec 2003.

70. Ryall, M. J. *Britannia Bridge: from concept to construction*, Proceedings of the Institution of Civil Engineers, Civil Engineering, **132**, May/Aug 1999.

71. Arya, C. and Vassie, P. *Whole Life Cost Analysis in Concrete Bridge Tender Evaluation*, Proceedings of the Institution of Civil Engineers, Bridge Engineering, **159**, Mar 2004.

72. Troitsky, M. *Prestressed Steel Bridges*, *Design and Theory*, Van Nostrand Reinhold, New York, 1990.

73. Zuou, P. and Zhu, Z. Concrete filled tubular arch bridges in China. *Structural Engineering International*, 3/97.

74. Highways Agency, *Design Manual for Roads and Bridges*, Volume 2, BD 67, Enclosure of bridges, HMSO, London, Aug 1996.

75. Calzon, J. *The Abacus System for the Launching of Large Span Constant Depth Composite Bridges – Types and Possibilities*, *Composite bridges – state of the art in technology and analysis*, Proceedings of the 3rd International Meeting, Madrid, Jan 2001.

76. Brown, C. *Reducing Noise Emission from Steel Railway Bridges*, Technical Report 173, Steel Construction Institute, 1997.

77. Steele, K., Cole, G., Parke, G., Clarke, B. and Harding, J. *Highway Bridges and Environment-sustainable Perspectives*, Proceedings of the Institution of Civil Engineers, Civil Engineering, **156**, Nov 2003.

78. Vetruvius, *De Architectura*, Harvard University Press, 1988.

79. Collings, D. Personal sketchbook, discussion at ABK architects, 1994.

80. Sibly, P. and Walker, A. *Structural Accidents and Their Causes*, Proceedings of the Institution of Civil Engineers, Part 1, **62**, May 1977.

81. Merrison, A. W. *Inquiry into the Basis of Design and Method of Erection of Steel Box Girder Bridges*, Interim report of the Committee on steel box girder bridges, 1981.

82. Beales, C. *Assessment of Trough to Crossbeam Connections in Orthotropic Steel Bridge Decks*, Transport and Road Research Laboratory, Crowthorne, Research Report 276, 1990.

83. Horne, M. *Structural Action in Steel Box Girders*, CIRIA Guide 3, London, April 1977.

84. Subedi, N. Double skin steel–concrete composite beam elements: experimental testing, *The Structural Engineer*, Nov 2003.

85. Johnson, P. *CTRL Section 1: environmental management during construction*, Proceedings of the Institution of Civil Engineers, Civil Engineering, special issue, **156**, Nov 2003.

86. Cooper, J. H. and Harrison, M. F. *Development of an Alternative Design for the West Rail Viaducts*, Proceedings of the Institution of Civil Engineers, Transport Engineering, **1536**, May 2002.

87. Forsberg, T. *The Øresund Approach Bridges, Composite bridges – state of the art in technology and analysis*, Proceedings of the 3rd International Meeting, Madrid, Jan 2001.

88. Gimsing, N. *Composite Action and High Strength Steel in the Øresund Bridge, Composite bridges – state of the art in technology and analysis*, Proceedings of the 3rd International Meeting, Madrid, Jan 2001.

89. Collings, D. *Design of Innovative Concrete Bridges for South China*, Bridge Management 3, E & F N Spon, London, 1996.

90. Dupre, J. *Bridges*, Konemann, Koln, 1998.

91. Tianjian, J. Concepts for designing stiffer structures, *The Structural Engineer*, Nov 2003.

92. Plu, B. *TGV Mediterranean Railway Bridges, Composite bridges – state of the art in technology and analysis*, Proceedings of the 3rd International Meeting, Madrid, Jan 2001.

93. Detandt, H. and Couchard, I. The Hammerbruke Viaduct, Belgium, *Structural Engineering International*, **13**, 2003.

94. Irwin, P., Mizon, D., Maury, Y. and Schmitt, J. History of the aerodynamic investigations for the Second Severn crossing, Proceedings of International Conference, Deauville, Oct 1994.

95. Highways Agency, *Design Manual for Roads and Bridges*, Volume 2, BA 61, The assessment of composite highway bridges, HMSO, London, Nov 1996.

96. Cracknell, D. *The Runnymede Bridge*, Proceedings of the Institution of Civil Engineers, May–Aug 1963.

97. O'Connor, C. *Roman Bridges*, Cambridge University Press, 1993.

98. Lutyens, M. *Edwin Lutyens*, Black Swan Books, London, 1991.

99. Benaim, R., Smyth, W. and Philpot, D. The new Runnymede Bridge, *The Structural Engineer*, **58**, Jan 1980.

100. Highways Agency, *The Appearance of Bridges and Other Highway Structures*, HMSO, London, 1996.

101. Yan, G. and Yang, Z. Wanxian Yangtze Bridge, China, *Structural Engineering International*, 1997.

102. Zang, Z., Pan, S. and Huang, C. *Design and Construction of Tongwamen Bridge, China*, Proceedings of the Institution of Civil Engineers, Bridge Engineering, **157**, March 2004.

103. King, C. and Brown, D. *Design of Curved Steel*, Publication P281, Steel Construction Institute, 2001.

104. Wardenier, J., Kurobanne, Y., Packer, J. A., Dutta, D. and Yeomans, N. *Design Guide for Circular Hollow Sections Under Predominantly Static Loading*, CIDECT, Ascot, 1991.

105. Bergmann, R., Matsui, C., Meinsma, C. and Dutta, D. *Design Guide for Concrete Filled Hollow Section Columns under Static and Seismic Loading*, Verlag TUV Rheinland, 1995.

106. European Stainless Steel Development Association, *Design Manual for Structural Stainless Steel*, 2nd edition, Steel Construction Institute, London, 2002.

107. Morris, R. *Notes on Sculpture, Art in Theory 1900–1990*, Blackwell, Oxford, 1992.

108. Scott, Q. and Miller, H. S. *The Eads Bridge*, University of Missouri Press, 1979.

109. Paxton, R. (Ed), *100 Years of the Forth Bridge*, Thomas Telford, London, 1990.

110. Bennett, D. *The Architecture of Bridge Design*, Thomas Telford, London, 1997.

111. Sharp, D. (Ed), *Santiago Calatrava*, Book Art, London, 1992.

112. Melbourne, C. (Ed), *Arch Bridges*, Proceedings of the First Conference on Arch Bridges, Bolton, 3–6 Sept, Thomas Telford, Sheffield, 1995.

113. Wurth, G. and Koop, M. Enneus Heerma Bridge, Ijburg, the Netherlands, *Structural Engineering International*, 2003.

114. Hesselink, B. and Meersma, H. Railway Bridge across Dintel Harbour, Rotterdam, the Netherlands, *Structural Engineering International*, 2003.

115. Friot, D. and Bellier, G. *Bonpas Tied Arch TGV Méditerranée High Speed Rail Bridge Over the A7 Motorway Toll Plaza, Composite bridges – state of the art in technology and analysis*, Proceedings of the 3rd International Meeting, Madrid, Jan 2001.

Steel–concrete composite bridges

116. Various authors. Second Severn Crossing, Proceedings of the Institution of Civil Engineers, **120**, Special issue 2, 1997.
117. Withycome, S., Firth, I. and Barker, C. *The Design of the Stonecutters Bridge – Hong Kong, Current and Future Trends in Bridge Design Construction and Maintenance*, Proceedings of the International Conference, Hong Kong, April 2002, Thomas Telford.
118. Collings, D. and Brown, P. The construction of Taney Bridge, Ireland, Proceedings of the Institution of Civil Engineers, Bridge Engineering, **156**, Sept 2003.
119. Iley, P. All set for the Olympic flame, *Concrete Engineering*, **8**, 2004.
120. De Miranda, F. *Design – Long Span Bridges*, ECCS/BCSA International Symposium on Steel Bridges, Feb 1988.
121. Concrete Society, *Design Guidance for High Strength Concrete, Technical report 49*.
122. Chatterjee, S. *Strengthening and Refurbishment of Severn Crossing*, Proceedings of the Institution of Civil Engineers, Structures and Buildings, **94**, Feb 1992.
123. Highways Agency, *Design Manual for Roads and Bridges*, Volume 1, BD 49, Design rules for aerodynamic effects on bridges, HMSO, London, May 2001.
124. Macdonald, J., Irwin, P. and Fletcher, M. *Vortex-induced Vibrations of the Second Severn Crossing Cable-stayed Bridge: full-scale and wind tunnel measurements*, Proceedings of the Institution of Civil Engineers, Structures and Buildings, **152**, 2002.
125. Maury, Y., MacFarlane, J., Mizon, D. and Yeoward, A. Some aspects of the design of Second Severn Crossing cable stayed bridge, Proceedings of International Conference, Deauville, October 12–15 1994.
126. Kumarasena, S., McCabe, R., Zoli, T. and Pate, D. Bunker Hill Bridge – Boston, *Structural Engineering International*, **13**, No. 2, May 2003.
127. Troyano, L. *Bridge Engineering a Global Perspective*, Thomas Telford, London, 2004.
128. Benaim, R., Brennan, M. G., Collings, D. and Leung, L. *The Design of the Pearl River Bridges on the Guangzhou Ring Road*, FIP Symposium on Post Tensioned Concrete Structures, London, 1996.
129. Russell, H. (Ed), *Bridge Design and Engineering*, Hemming, 2004.
130. Maguire, R. and Mathews, P. *The Iron Bridge at Coalbrookdale*, (publisher unknown) 1958.
131. Network Rail, *Railtrack Line Code of Practice*, RT/CE/C/025, Feb 2001.
132. Highways Agency, *Design Manual for Roads and Bridges*, Volume 3, BD 61, The assessment of composite highway bridges and structures, HMSO, London, Nov 1996.
133. Chrimes, M. *The Development of Concrete Bridges in the British Isles Prior to 1940*. Proceedings of the Institution of Civil Engineers, Structures and Buildings, **116**, Aug–Nov 1996.
134. Concrete Society, *Durable Post Tensioned Concrete Bridges*, Technical Report 47, 2nd edition, 2002.
135. Johnson, R. and Cafolla, J. *Corrugated Webs in Plate Girder Bridges*. Proceedings of the Institution of Civil Engineers, Structures and Buildings, **123**, May 1997.
136. Johnson, R. *Analyses of a Composite Bowstring Truss with Tension Stiffening*, Proceedings of the Institution of Civil Engineers, Bridge Engineering, **156**, June 2003.
137. Richardson, J. *The Development of the Concept of the Twin Suspension Bridge*, National Maritime Institute, Oct 1991.
138. Tucker, M. *Canal Bridge – Bishops Bridge Road, Preliminary Archaeological Report*, English Heritage, London, Report B/019/2003, Oct 2003.
139. Ryall, M., Parke, G. and Harding, J. (Eds), *Manual of Bridge Engineering*, Thomas Telford, London, 2000.
140. Menn, C. *Prestressed Concrete Bridges*, Birkhauser Verlag, London, 1990.

Index

Page numbers for figures and illustrations are shown in italics. Bridges cited are in the UK unless otherwise stated.